Tribute
to a
Mathemagician

Tribute
to a
Mathemagician

Edited by

Barry Cipra
Erik D. Demaine
Martin L. Demaine
Tom Rodgers

A K Peters
Wellesley, Massachusetts

Editorial, Sales, and Customer Service Office

A K Peters, Ltd.
888 Worcester Street, Suite 230
Wellesley, MA 02482
www.akpeters.com

Library of Congress Cataloging-in-Publication Data

Tribute to a mathemagician / edited by Barry Cipra ... [et al.].
    p. cm.
    Includes bibliographical references.
    ISBN 1-56881-204-3
    1. Mathematical recreations. I. Cipra, Barry.

QA95.T744 2004
793.74--dc22

                    2004053370

Printed in Canada
09 08 07 06 05   10 9 8 7 6 5 4 3 2 1

Contents

Part III   Brainteasers   55

Part IV   Braintempters   83

Preface

This book is the third in a series devoted to Martin Gardner. Each book is a collection of articles written by a handful of the many mathematicians, puzzlers, and magicians that Gardner inspired. Most of these articles were presented at the *Gathering for Gardner (G4G)* meetings, a regular gathering of enthusiasts who share some of Gardner's many interests. This volume draws on the many excellent presentations at the last such meeting, G4G5, held in Atlanta on April 5–7, 2004. The previous volume in the series, *Puzzlers' Tribute: A Feast for the Mind* (2002), is based on three of the Gatherings—G4G2, G4G3, and G4G4. The first volume, *The Mathemagician and Pied Puzzler* (1999), is based on the first Gathering, G4G1. At the back of this book, you will find an index to all three books in the series.

One of the running traditions of the Gatherings is to have a theme related to the number of the conference. The theme for G4G5 was "penta"— Greek for the number five, as in "pentagon" and "pentagram." This theme shows up in several of the articles in this book.

Martin Gardner is the father of recreational mathematics, an avid puzzler, a life-long magician, and a debunker of pseudoscience. He has written more than 65 books throughout science, mathematics, philosophy, literature, and conjuring. He has deeply influenced countless readers of his "Mathematical Games" column in *Scientific American*, which ran for 25 years from 1957 to 1982. This column popularized recreational mathematics, and introduced many connections between mathematics, puzzles, and magic. Together with Gardner's amazing ability to connect with his many readers, the columns gave the general public the opportunity to enjoy mathematics and to participate in mathematical research. Many of today's mathematicians entered the field because of Gardner's influence. A whole body of research into recreational mathematics has also emerged, solving problems that Gardner posed years ago and introducing new problems in the same spirit.

Although Gardner himself has unfortunately been unable to attend the Gatherings since G4G2, the presentations are video-recorded so that he can watch at home. Gardner's great gift to us was to interconnect so many people and so much beauty in each of his fields of interest. Our gift back to Gardner is to show that the community he built continues to flourish in his honor.

This book also commemorates the memory of two great puzzlers, who were good friends of Gardner. Edward Hordern was a master puzzle solver and a great puzzle collector. Nobuyuki Yoshigahara—affectionately called Nob—was a master puzzle designer who greatly influenced the entire puzzling community. The first four articles in this book recount stories of these two great men.

This book could not have happened without the help of countless people. The organizers of the Gatherings for Gardner—Elwyn Berlekamp, Tom Rodgers, and Mark Setteducati—would like to acknowledge the work of many people who have helped make the Gatherings successful, including Scott Kim, Jeremiah Farrell, Karen Farrell, Emily DeWitt Rodgers, David Singmaster, and many others. The book itself exists thanks only to the efforts of a large group of contributors. Scott Kim conceived of and assembled the first tribute volume, and designed the cover in this third volume and the second. Emily DeWitt Rodgers has done an excellent job of designing and maintaining the g4g4.com website, which is devoted to the Gatherings. We also thank a number of anonymous experts for their reviews of papers. Finally, the support of our publisher, A K Peters, Ltd., enabled this project to become the book you now hold in your hands.

All of us feel honored by this opportunity to join together in tribute to the man in whose name we gather, Martin Gardner.

Barry Cipra
Northfield, Minnesota

Erik D. Demaine
Cambridge, Massachusetts

Martin L. Demaine
Cambridge, Massachusetts

Tom Rodgers
Atlanta, Georgia

In Memoriam

Edward Hordern (1941–2000)

James Dalgety

In half a century Edward Hordern seemed to live several lifetimes worth of diverse activity. Certainly in his last quarter century he managed at least two lifetimes worth of puzzling.

After studying Modern Languages at the Sorbonne, University of Paris, Edward went into accountancy for two years. He then joined a large advertising agency, Alfred Pemberton Ltd and its subsidiary Business Press Bureau Ltd. in 1963. He spent five years with the firm, starting as a trainee and ending up a director. In 1966 he started his own company, Creteco Ltd, manufacturing and selling accessories for concrete. As a sideline to this business, he had twelve trucks doing long distance haulage—the longest trip on one occasion was from England to Bombay. His firm was also the first to import commercial maintenance-free batteries into Europe.

In 1970 Edward took over the running of the family farm at Cane End. The farm consisted of 1,200 acres (500 Hectares) of mixed cereal and beef. He planted 12 acres of vines. In one year of the International Wine and Spirit Competition, his 1986 vintage won a silver medal and his 1987 vintage won a bronze. This was achieved competing against 1100 entries from 30 countries. His wine label not only showed his beautiful 18th Century

James Dalgety Metagrobologist. Ex-Antarctic Weather Observer and Turkey Farmer(1960s). Co-Founder of Pentangle Puzzles (1970s & 80s), Project Director of Bristol Exploratory Hands-on-Science Centre (1980s), Exhibition Consultant to G4G1 Atlanta, Instigator & Designer of PuzzleQuest in Cardiff, and other related Hands-On Exhibitions. Curator of the Hordern-Dalgety Puzzle Museum. http://puzzlemuseum.org.

manor house, but also incorporated a maze in which one must find a path from the bunch of grapes to the bottle, along dark lines only.

Edward extended his many interests to include local politics. He was first elected as a Councillor to the Oxford County Council in 1985, and he was re-elected in 1989. His special brief was as spokesman for Museums, Libraries, and Recreation.

I first met Edward in 1973. He wrote to me, when I was co-founder and director of Pentangle, wanting to know the shortest solution to one of our sliding block puzzles. It transpired that his solution was already far shorter than we had thought possible. I phoned him up and asked him if he was a collector. He asked what I meant, so I asked him what he did with his solved puzzles. As he put his solved puzzles into a drawer instead of giving them away, I said he must be a collector. We soon visited each other and became great friends.

At this stage Edward's puzzling had mainly consisted of buying good wooden jigsaws of 1,000 to 2,000 pieces and an occasional cardboard jigsaw with up to 5,000 pieces. He had just discovered three-dimensional jigsaws and had commissioned a magnificent box of fifteen of Peter Stocken's beautiful puzzles. His mechanical puzzle collection, of around 200, fitted into a few drawers, and consisted primarily of Sliding Block puzzles, which he had found in books and made for himself out of Lego. As he got really enthusiastic about mechanical puzzles, the two-dimensional jigsaws were all put away in a cupboard. His house was a large one, and when I say "cupboard" I really mean a small attic room of several cubic meters. The two-dimensional jigsaws were still there filling the cupboard and gathering dust when the house was emptied in 2001. I am happy to say that all the fine wooden ones are now in The British Jigsaw Puzzle Library.

When Edward first visited my house, it was around the time that Jerry Slocum had told me that I had the world's third largest puzzle collection. Small by the standards of today's big collections, it nevertheless inspired Edward. He saw the huge variety that existed, and his enthusiasm knew no bounds. For the next 25 years he collected at an unsurpassed rate. He continued with all his extensive business activities and with his regular trips to Scotland, pheasant shooting. He seemed to effortlessly add puzzling full-time into his already full life. Instead of taking two holidays per year, one with his wife, and one with all the family, he added a third holiday for puzzling. He went to 19 annual "International Puzzle Parties" and, as they outgrew the small gatherings at Jerry Slocum's house, Edward joined Jerry and Nob Yoshigahara to form the supervising committee of the IPPs. He hosted the 10th and 19th IPPs in London. He only needed six hours sleep per night, and he is even reputed to have occasionally taken puzzles to bed.

Edward Hordern in Tokyo in 1995.

His enthusiasm for sequential puzzles culminated with the publication of his book *Sliding Piece Puzzles* in 1986 by Oxford University Press. When one considers that this book was published before computers were in general use, it was an incredible achievement. Despite developments since, it remains the standard text on the subject, with many of his hand-done solutions remaining unbeaten.

In 1993 Edward privately published a wonderful Centenary edition of Professor Hoffman's *Puzzles Old & New*. It was fully corrected and with the addition of colour photographs of the original Victorian puzzles from his and other private collections. As "senior proof-reader" to the project, Laurie Brokenshire can attest to the very large numbers of errors, in both the original puzzles and their solutions, that Edward managed to spot.

Meanwhile he was filling his large house. Initially it was just his study, and then he started to encroach on the dining room. Fortunately, in the reign of Queen Anne, manor houses were built with serious attic rooms so, without upsetting his wife too much, he took over one of these. However, by 1999 he had puzzles in his study, in half the dining room, and in three attic rooms that were specially fitted with display furniture. He was also using a further attic room and most of the extensive loft space as a dumping and sorting ground for puzzles.

Cane End Wine Label. Follow the dark lines from the grape icon in the middle left, through the bushes, across the lower ledge of the upper roof, through the bushes again, to the bottle icon in the middle right. (See Color Plate I.)

Edward collected all kinds of puzzles in an extraordinarily thorough way. Today it has become almost impossible to collect in the same manner. Few people have the space, and fewer are willing to learn about such obscure things as the care of 18th Century puzzle ephemera. Truly antique puzzles are becoming ever more difficult to find. The other big change in recent years has been the explosion of creativity in the puzzle world. This has generated such a profusion of new puzzles that it is difficult for anyone to collect all new puzzles as comprehensively as was possible in the past. This wonderful productivity is due to many factors, including Martin Gardner's writings and a few companies such as Pentangle in the UK, and Mag-Nif in the US, who, in the early 1970s, brought the puzzle market back from the oblivion it had suffered since the depression after the WW1. The International Puzzle Parties, the Gatherings for Gardner organised by Tom Rodgers, home computers, and above all the Internet have created a wonderful worldwide exchange of information. In the past, puzzlers were forever re-inventing the wheel, whereas now they progress from developments of others. Edward did not have much time for computers but he had a great influence on all the other areas of the puzzle world, and he is greatly missed by all who knew him.

Edward built up the finest puzzle collection in the world. To do this one needs to be a bit obsessive, but he never allowed his obsession to rule him. He remained a gentleman, interested and knowledgeable in a wide range of subjects quite unrelated to puzzles. He did not just collect puzzles. He solved his puzzles. He shared them with other people and always welcomed fellow puzzlers at his house. He was very generous in exchanging puzzles. It was wonderful that on his last antique puzzle hunt in London, only two weeks before his death, he found the last puzzle that he needed to complete his favourite group—his wooden Hoffmann puzzles.

Edward died of cancer on May 2, 2000 at the age of only 58. Throughout his painful illness he remained uncomplaining and continued collecting puzzles. His wife Wendy, two sons, and two daughters survived him. He knew that his first grandchild was expected soon, but very sadly did not live to meet his grandson Barnaby, who was born in June 2000.

I was overwhelmed when Edward's family gave me his collection. It was extremely generous of them. Even though I had packed the most delicate items myself, it took weeks of work on the part of a professional removal firm to move it. It was as if Edward had given me the ultimate sliding block puzzle. I have now had the collection for nearly two years and have amalgamated my own with it and separated out the duplicates. I still have not found time to finish counting it, let alone catalogue it all. I enjoy puzzling every day and the only thing that spoils it is that Edward is not around to enjoy it as well. I hope I can do justice to the collection and find a way to finance providing it with a permanent home within the UK before too many years pass. This would be a great memorial to a great collector.

Edward Hordern—Puzzle Solver Extraordinaire

Dick Hess

I first met Edward Hordern at the Second International Puzzle Party at Jerry Slocum's house in April 1979. I then visited him in England in July of that year at his house in Wyfold Grange. His impressive collection was taking good shape by then, but more impressive to me was his dedication as an enthusiastic puzzle solver. He had a rule: "A puzzle doesn't go on the shelf until I've solved it." Clearly he'd solved a lot of puzzles, because his shelves and drawers were packed with wonderful items. At that stage he was the world's expert at solving sliding block puzzles and had recently produced a 65-page book, *150 Sliding Block Puzzles with Solutions.* His solutions were all done by hand, taking hours of painstaking effort and resulting in careful notations of the solutions in his own hand. He was always eager to find improvements to what he had already developed. The effort was later expanded and published as his classic *Sliding Piece Puzzles* in 1986 with Oxford University Press.

An impressive display of puzzle rings, many made by José Grant, graced his wall. Amazingly, he had mastered the solutions to all of them. He explained to me that the rings with an odd number of bands (7, 9, 11, ...) proved the most challenging. His aptitude at solving these puzzles

Dick Hess is a long-time enthusiast of recreational mathematics and mechanical puzzles.

demonstrated his superior ability to think in three dimensions. He had also been a regular customer of Stewart Coffin for several years at that time. He told me that he asked Stewart to send him puzzles disassembled so that he could enjoy the challenge of solving them without the clue of seeing their final form. To my knowledge Edward always rose to this challenge and solved every one of Stewart Coffin's puzzles.

The Rubik's Cube craze was just hitting in 1979. Over the next few years Edward spent many happy hours solving Rubik's Cube and the avalanche of variants it inspired. He documented his efforts for each puzzle and in 1986 produced a booklet, *Solutions to Various Rotational Puzzles and Mechanical Sliding Piece Puzzles*, which gave his approach to solving no less than 21 puzzles. These included the Hungarian Rings, Engel's Enigma, Trillion, Ten Billion, Skewb, Dodecahedron and Rubik's Magic. As more such puzzles hit the market, Edward would solve each and carefully document his findings. Among these were Hungarian Stop Watch (1983), New Fifteen (1984), Rubik's Master-Magic (1987), Topspin (1989), and Roundy (1990).

Edward visited me in 1999 during one of his visits to Los Angeles for medical treatment. His enthusiasm for puzzles and puzzle solving was not reduced a bit by his medical condition, and as usual he asked if I had anything new in puzzling. I offered him two variants of the bent nails puzzle recently produced in Australia. He took them with him to bed and by morning had not only solved each but had joined one piece from each to create an entirely new and even better puzzle with two separate levels of difficulty. Edward was truly a puzzle solver extraordinaire. He is sorely missed by those who knew him.

Nob Yoshigahara

Jerry Slocum

Figure 1. Nob at AGPC Convention, 2003.

Puzzlemaster, genius, International Ambassador of Puzzles, legend, and one of the best puzzle inventors that the world has ever known are some of the terms that have been used to describe Nob. His enormous accomplish-

Jerry Slocum is the author of nine books about mechanical puzzles and is also known for his research on the history of puzzles and his large collection of puzzles and puzzle books. He is the founder and organizer of the annual International Puzzle Parties, held in the United States, Europe and Asia.

Figure 2. Nob's AGCP double silhouette badge.

ments encompass inventing and designing about two hundred mechanical puzzles, which were produced commercially, with at least eight million sold in Japan and America. Rush Hour, made and sold by ThinkFun, is one of the most popular puzzles of all time. It has received fourteen awards from educational and popular publications. Nob also invented or helped to develop other ThinkFun puzzles: Hoppers, FlipIt, Shape by Shape, and Lunar Lockout.

At its annual convention in 2003, the Association of Game and Puzzle Collectors (AGPC) awarded their highest honor, the Sam Loyd Award, to Nob Yoshigahara (see Figure 1) for his "Lifetime achievements in the design of mechanical puzzles." The convention badge (see Figure 2) shows a pair of left and right facing silhouettes of Nob forming the stand supporting several of his puzzles. His acceptance of the award is shown in Figure 3.

Figure 3. Nob accepts the AGPC Sam Loyd award for his lifetime achievements in puzzle design.

Figure 4. Nob's first mechanical puzzle, at age 19, the DuaLock.

Mechanical Puzzles

Nob invented his first mechanical puzzle in 1955 at age nineteen. He named the puzzle DuaLock, which provided a clue to the solution. Figure 4 shows a transparent version and the wooden production version made by Japanese puzzlemaker, Hikimi. To solve the puzzle, you must spin it to release one set of pins that lock it together. Then you must carefully turn it over, while preventing the first set of pins from re-locking the pieces, and spin it a second time to release the second set of pins.

During the last twenty-five years, Nob has been a driving force in developing, promoting, and popularizing puzzles in Japan. He designed or helped develop almost all of the forty-nine unusual and elegant pieces of the Glass Puzzle Collection from Toyo Glass Company. The Pack the Plums puzzle (see Figure 5) was their best seller. The plums represent extraordi-

Figure 5. Pack the Plums, in front, was the best selling of the Toyo Glass puzzles.

Figure 6. Two Hikimi wooden puzzles, with Nob's T puzzle on top.

narily sour Japanese pickled plums, and if you succeed in packing them in
the glass—a very difficult task—you can turn the glass upside down, and
the nine pieces will not fall out.

Nob was invited by the Mayor of Hikimi, a small Japanese town that
was losing population, to supervise the establishment of a large puzzle
exhibition and Puzzland Hikimi, a factory that made wooden puzzles using
wood from their nearby forests. Nob also designed many of the fifty-seven
wooden puzzles made by Hikimi. Figure 6 shows two of the Hikimi puzzles.
Many of Nob's puzzles produced by Hikimi are shown in Nob's book, *Puzzle
in Wood,* published in Japan in 1987 by Tokuda.

Hanayama's thirty-five-piece Cast Puzzle series was also developed by
Nob. The series began using Nob's improved versions of antique cast iron
puzzles from the USA and England. Then Nob developed numerous new
cast puzzles. Some were based on ideas from expired patents but were
considerably improved to provide excellent puzzles. He also worked with
other puzzle designers from Japan, Europe, and America to develop many
new cast puzzles. Figure 7 shows the improved, classic Star puzzle, Flag
puzzle (based on an expired American patent), and Akio Yamamoto's new
Amour puzzle. Nob's television appearances and continuous video sales
pitch that played in 3,000 stores throughout Japan has increased the sales
of Hanayama's cast puzzles ten-fold in the last several years.

Japanese customers have found that his silhouette on the boxes of this
series of puzzles (see Figure 8) guarantees a very high quality puzzle that
is fun to solve. And his puzzles have developed a large following. This is
shown by the interesting fact that the most difficult cast puzzles are the
best sellers.

Figure 7. Cast Puzzles by Hanayama: Flag, Star, and Amour.

Figure 8. Nob's silhouette on a puzzle indicated that it was of high quality and fun to solve.

Nob had an extraordinary talent for improving old puzzle designs as well as inventing entirely new puzzles. While he was translating Jerry Slocum's *New Book of Puzzles* (W H Freeman & Co., 1992) into Japanese, he realized that one of the solitaire puzzles, The Great 13 Puzzle, patented in 1899, had a lot more potential than was utilized in the original puzzle. He and his team of Japanese designers came up with 40 problems, in four levels of difficulty, using the same board. This puzzle was manufactured and sold by ThinkFun under the name Hoppers. When Hoppers came out, my grandchildren, Jack (age 7) and his sister, Sydney (age 11), were so taken by the puzzle that they sat down for several hours without getting out of their chairs and solved all 40 problems that came with the puzzle. (See Figure 9.)

Another example of Nob's ability to redesign an old puzzle is the 100 year old classic "T" Puzzle. Nob modified the dimensions of the four pieces of the T Puzzle and provided twenty additional assembly problems to be solved with the new pieces. More than four million copies of Nob's puzzle The-T have been sold (see Figure 6).

Nob has helped and taught many puzzle designers in Japan, Europe, and America and has helped get puzzles of other designers produced. Nob's studio in Tokyo was a friendly gathering place for puzzlers from around the world, with thousands of puzzles with which to play. Figure 10 shows Nob

Figure 9. The original 13 Puzzle (left) and Nob's improved version, Hoppers, by ThinkFun (right).

Figure 10. Nob with puzzle designers Oskar van Deventer and Bill Cutler.

Figure 11. Puzzle designer and craftsman Akio Kamei and Master Craftsman Yoshiyuki Ninomiya with Nob.

with two other famous puzzle designers with whom he worked, Oskar van Deventer from the Netherlands and Bill Cutler from America. Figure 11 shows Nob with his friends, famous Japanese puzzle designer and craftsman Akio Kamei and Master Craftsman Yoshiyuki Ninomiya.

Figure 12. Puzzle collectors the late Edward Hordern, Nob, and Jerry Slocum.

Owning 14,000 puzzles, Nob had the largest collection of puzzles in Asia, as well as a huge library of puzzle books. Figure 12 shows him with two puzzle collector friends, the late Edward Hordern and Jerry Slocum.

Math Puzzles

Nob also created mathematical puzzles too numerous to count, wrote sixty puzzle books, produced as many as sixteen monthly columns, and translated English and German puzzle books into Japanese. Three longtime American friends who enjoyed challenging Nob, and vice-versa, on math puzzles are Don Knuth, Sol Golomb, and Dick Hess. The next article includes reminiscence from them.

Puzzle Parties

One of Nob's favorite magic tricks was making a ball of crumpled paper disappear in front of some unsuspecting subject. Figure 13 shows Nob just after he made the paper ball disappear for Ellen Ireland at IPP 7. Everyone in the room except Ellen saw Nob throw the paper ball (shown above her head) from his right hand an instant before the photo was taken. He also loved to entertain with his slight-of-hand skills using coins, by making them penetrate table tops and magically appear from the ears of amazed, and adoring, children.

In 1982 he brought his son, Takayuki, to his first International Puzzle Party (IPP 5), and Nob never missed one since. Larry Nichols, inventor of

Figure 13. Nob's magic made the paper ball disappear for Ellen Ireland.

Figure 14. Nob-in-the-box makes a spectacular entrance at IPP 7.

a puzzle similar to the Rubik's Cube, attended IPP 7 in Beverly Hills in 1984. Nob burst into the Party from his hiding place, a cardboard Rubik's Cube box (see Figure 14).

Nob was the host of the first International Puzzle Party held outside America: IPP 9 in Tokyo in 1988. He also was the host of the next two IPPs held in Japan: IPP 12 (1992) and IPP 15 (1995). A special host gift, designed and made by Gary Foshee, was given to Nob at IPP 15 as a token of our thanks. Figure 15 shows Nob's gift (his famous silhouette in Lucite and a six-piece Burr) being handed from Akio Kamei to Jerry

Figure 15. Nob is awarded a special trophy at IPP 15 for hosting three Japanese IPPs.

Figure 16. Poster for the Puzzles Old and New Exhibition at the Matsuya Ginza department store.

Slocum, IPP Founder, who presented the award to Nob. Nob also helped organize and contributed to the Puzzles Old and New Exhibition at the Matsuya Ginza department store on the Ginza in Tokyo, which attracted 50,000 paid admissions in two weeks in April 1988. The poster for the exhibition is shown in Figure 16.

He was proud of his family and often brought his wife, Takako, or one of his children with him to Puzzle Parties and on his frequent trips to Europe and America. Figure 17 shows his wife, Takako; daughter, Chisato; and grandson, Shun; attending IPP 21 in Tokyo in 2001.

Figure 17. Shun, Chisato, and Takako, Nob's grandson, daughter, and wife.

Nob's Life

Nob did not have an easy life. At age seven, he left Tokyo because of the bombing danger during World War II and was living with his grandfather in Iwakuni, near Hiroshima. He found an English Solitaire board in the dusty attic of the house and solved the very difficult puzzle of reducing the pegs to one by jumping. When he showed his solution to the adults, who could not solve it, his maid considered him a child prodigy or Second Buddha. He wrote in 1990 that he still recalls the flash of light and sounds when the atomic bomb went off in Hiroshima. When he returned to Tokyo, he says, "I found it all burnt." Here is Nob's interpretation of these events: "Without this evacuation, exaggeratedly saying, without World War II, my puzzle life now won't exist. The war deliberately brought me a chance to go from puzzle, to puzzler to puzzlest." A chemical explosion burned him badly while he was in his late twenties, and during the last ten years of his life, he had serious problems eating, because his stomach had been removed due to cancer.

However, one little known fact about Nob's ancestry can help explain his drive for excellence, his persistence, and his philosophy of never giving up. He is descended from an Edo era Samurai who served the Mouri Family, a feudal lord of the Chugoku region of Japan. Nob, however, used his brain, rather than his sword, to solve Gordian Knot puzzles.

He is quite famous for his impossible puzzles, magic, jokes, wit, puns in English, and sense of humor. One of his mechanical puzzle jokes is what appears to be a six piece burr that Nob has carved out of a solid piece of wood. "Nob's Three-piece" burr—commonly made out of three notched sticks of wood which are orthogonally interlocked—he made out of two notched pieces of wood, as shown in Figure 18.

On June 11, 2004, a week before he died in his studio in Tokyo, Nob visited our home in Beverly Hills. My son, Allan, and grandson, Jack,

Figure 18. Nob's three-piece burr is assembled from 2 pieces.

Figure 19. Nob entertaining with an electric guitar.

stopped by to visit with Nob. Both had known Nob for many years. Jack brought his new electric guitar with him to show his grandparents and Nob. After Jack demonstrated some cords and melodies, Nob was asked if he would like to try playing the guitar. Much to our surprise he said "sure" and proceeded to play cords and melodies (see Figure 19). When we asked him how he knew how to play the instrument, he said that he had played the ukulele in a Hawaiian band. Jack also showed Nob the American way to eat an Oreo cookie: twist and lick! After we heard that Nob had died, I asked Jack what he thought about Nob. Jack said Nob was "Awesome . . . Totally Cool!"

Last year in the "Biographical Information" section of the *Directory of Puzzle Collectors and Sellers*, Nob wrote, "Petroleum engineer, teacher of chemistry, retired, now. I want to be invited to any place in the world. It is my **MUST!**"

And travel he did. Legend has it that if more than six puzzlers got together anywhere in the world, Nob would show up. He was always more than welcome and very generous with his time, his attention, and his gifts of puzzles.

As we miss Nob, let us celebrate his amazing life, his contributions to the world of puzzles, and most of all, his friendship. Nob would want to say "Goodbye" with his appropriate tag line, "Happy Puzzling Forever!"

Nobuyuki Yoshigahara
(1936-2004)

Nob deeply impacted the lives of many puzzlers. Assembled here are brief personal recounts from a handful of his friends and collaborators.

Nob's letters to me were always inspiring, and they would often cause me to drop everything else I was doing; the problems he posed were so fascinating, I simply couldn't continue work-as-usual. During the summer of 1994 we were sending letters to each other more than once per week, so that they would cross in the mail.

My favorite reminiscence is about the time I sent him my "Pentagon Puzzle Challenge" in September 1994; the problem was to take four small pentagons (red, white, blue, green) and cut them each into four golden triangles—namely, isosceles triangles where the unequal sides have ratio phi to each other (the golden ratio)—then to reassemble those triangles into a larger pentagon, with no two pieces of the same color touching. I had needed a computer to solve it; and I knew that the solution was unique. Early in October he sent a postcard containing an elegant "hiroi-mono" puzzle and also added the following note: "I solved your extremely difficult problem in only 20 minutes, but some mania puzzler consumed 10 days in vain!" A few days later I received his letter with the detailed answer; and then came the biggest surprise of all: The postman brought me a package containing a beautiful model of the puzzle, handcrafted from five

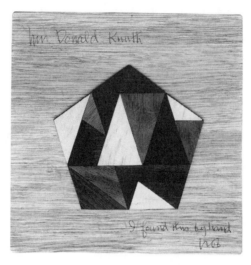

Handcrafted solution to "Pentagon Puzzle Challenge" autographed by Nob.

kinds of wood—exhibiting his solution, and autographed too! This wonderful artifact remains the most cherished item in my personal collection of puzzles.

—*Don Knuth*

I first met Nobuyuki Yoshigahara at one of the early IPP's (International Puzzle Parties) when they were still held at Jerry Slocum's house. Nob told me that it was my work on polyominoes that first got him interested in puzzles. I'm sure he was also quite amused by my attempt to speak Japanese, but was polite enough not to laugh.

Over the years we met on innumerable puzzle-related occasions, and often challenged each other with new puzzles by mail. In early July 2003, while attending a symposium in Yokohama, I used the free Wednesday to take an early train into Tokyo where I met Nob at his studio. He invited several other prominent puzzlists to join us, and we spent a fascinating day together. This included a visit to an outdoor puzzle vendor whose large display of geometric puzzles had many invented by Nob and his friends. He said that as the "famous inventor of pentominoes" I was entitled to a discount, and I bought more puzzles than I could easily carry. After lunch Nob led us to the Tokyo Science Museum, which was running a special exhibit on remarkable mechanical devices (several of them puzzle-like) from the Tokugawa era. After dinner together, it was quite late when I finally returned to my hotel in Yokohama.

A month later I saw Nob again (for what proved to be the last time) at the IPP in Chicago. After that, we continued to exchange emails, including some comments I had about Nob's book *Puzzles 101* (A K Peters, 2004), which I have in both Japanese (a gift from Nob) and in English. I was abroad when he visited California in June 2004, and I was shocked and deeply saddened to learn of his death when I returned.

Nob's creativity in inventing, improving, collecting, and describing puzzles of all sorts was truly unique. His impact on the world of puzzledom was profound, and his influence will be enduring.

—Sol Golumb

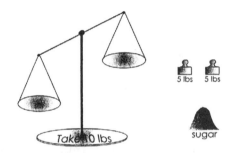

Here is a scale with uneven arms.
Using two 5 pound weights, measure exactly 10 pounds of sugar.

When Nob was only ten years old, he created the problem shown above, sent it to a newspaper, and won a Grand Prize. The puzzle asks to measure exactly 10 pounds of sugar, given two five pound weights and a scale. The catch is that the scale is uneven: without any weight on either plate, one plate is higher than the other.

The solution: Put ten pounds of weight (5 lbs + 5 lbs) on a plate, and put sugar on another plate until it balances. Then, remove the weights, and place sugar there until it again balances. The sugar is exactly ten pounds. This problem is in Nob's new book, *Puzzles 101*, published by A K Peters in 2004.

Nob thoroughly enjoyed solving math puzzles, as well as creating them. Nob worked very hard one night on "Problem Number 2281, Ten Tough Trapezoid Tiling Tasks," by Bob Wainwright, in Volume 27(4) of the *Journal of Recreational Mathematics*. The problem is this: "Starting with a regular trapezoid (top and sides each equal to half the base length), determine all possible dissections which will result in similar parts." Ten specific tasks are then defined. Nob was so excited at finding an infinite set of solutions, he phoned me after midnight to tell me of his accomplishment, and later wrote, "I nearly went out of my mind when I discovered it."

—*Dick Hess*

I have long considered Nob Yoshigahara to be one of the great puzzle designers. But much of his work is not well known outside of Japan. So when a copy of his first English language book *Puzzles 101: A Puzzlemaster's Challenge* (A K Peters, 2004) appeared in my mailbox, I jumped on the chance to feature his work in my monthly Bogglers puzzle column in *Discover* magazine (volume 25, number 8, August 2004). *Discover* is a science magazine, so I took the liberty of embellishing his puzzles with a science theme. I also added easier variations of each puzzle to ease readers into each challenge. I showed a draft of this article to Nob when I visited him in San Francisco, shortly before his death. Nob enjoyed collaborating with other puzzle designers, so as I expected, he had no problem with my embellishments.

—*Scott Kim*

Nob dragged many people into puzzles. I am one of them. About 12 years ago, I got a national grant to visit Europe and the United States to study my academic major. Before the visit, I phoned Nob Yoshigahara. He told me about several puzzle people and joyfully recommended that I visit them. And so I met Matti Linkola in Kajaani, David Singmaster in London, James Dalgety in Manmead, Bob Easter in San Francisco, and Harry Nelson in Livermore. Meeting all of them, I felt happy and was much shocked. Devoted to collecting puzzles as a natural result, I began to join International Puzzle Parties (IPP). The first one I attended was in Culver City, and since then I have attended every IPP. I was greatly honored to host IPP 21 in 2001.

Nob did not drive a car. He took young puzzle people with him to act as his drivers when he traveled. As a result, they experienced much in the puzzle community. I drove for the first time on the right side of the road when driving with him to San Diego in 1991 on the way to meet Harry Eng.

Nob hated sports and quizzes. His opinion was that puzzles can be solved only by thinking without memorized knowledge, but such knowledge is necessary for quizzes. However, his power of memorization was great, except when he was with alcohol.

I remember several practical jokes that Nob played. He made a trick out of the standard puzzle of two bent nails. He bent one of the nails in

Nob at the secondary gathering with a drink after a monthly Academy of
Recreational Mathematics, Japan (ARMJ) meeting, 1998.

the reverse way, so that the puzzle could not be taken apart. He also made
an object like a 3-piece burr, but it was made of a single piece of wood.

Nob was fond of the phone number of his studio, 3267-7623, because
it was a palindrome. I pointed out the longer palindromic sequence 1089,
2178, 3267, 4356, 5445, 6534, 7623, 8712, 9801. Each four-digit number is
obtained from the previous by adding 1 to the first two digits and subtract-
ing 1 from the last two digits. Among the four-digit numbers are Nob's
3267-7623, making this number much more than just a palindrome. He
was very glad. Sadly, we neither visit his studio nor phone that number
any more.

—Yoshiyuki Kotani

Nob helped me with many puzzles, and greatly aided my site and columns.
When I was unemployed, and didn't do well at a puzzle convention, he
bought all my unsold puzzles and took them to Tokyo (where he sold them
all). He got my puzzles published in Japan for me, and got me paid for
them. I am just one of many hundreds of puzzlemakers he has helped.

One of Nob's most famous puzzles in North America is Rush Hour. In
a posting to the Nobnet mailing of puzzle enthusiasts, Nob adds a new
variation on this puzzle:

> According to normal RUSH HOUR rules, your car XX must go
> out from this garage through the exit (=>). Only one exception
> for this problem is, you can remove one car. Which car should
> be removed to let your car go out? Do not remove the XX itself.
> Find its shortest moves.

```
+  -   -   -   -   -   -   +
|  O   P   P   P   A   A   |
|  O   B   C   C   D   D   |
|  O   B   X   X   E   F   =>   exit
|  G   H   I   I   E   F   |
|  G   H   Q   Q   Q   J   |
|  K   K   R   R   R   J   |
+  -   -   -   -   -   -   +
```

Nob long had frail health, but he relentlessly promoted puzzles right until the end. In the hours before going into major surgery, for example, he programmed a computer to solve a complex pentomino-packing problem. He didn't want his hours of unconsciousness wasted. His humor, wit, and force of will made him seem immortal. I will greatly miss him.

—Ed Pegg Jr.

The regular monthly meeting of the Academy of Recreational Mathematics, Japan (ARMJ) is held every third Saturday, and Nob regularly attended the meetings. When he did not appear at the meeting on June 19, 2004 (the third Saturday of June), some members visited Nob's studio after the meeting. They found him deceased on the side of his desk with a peaceful expression on his face. His computer was still on. No special cause of death was reported by the official inspector. Nob had just returned from the west coast of the USA that Thursday after meeting with some puzzle friends there. Most of his puzzle friends, including those who met with him just a few days prior in the USA, were surprised and grieved over his unexpected, sudden departure.

A field that interested Nob besides puzzles was magic. He was a member of TAMC (Tokyo Amateur Magicians Club), and he sometimes performed card magic for us using his professional skills.

Almost all of his friends recognized that he had a great sense of humor and was fond of jokes. Last summer, someone gave him a New York Yankees baseball cap. He was delighted to wear it even though he does not play sports, because the "NY" logo on the cap matches his initials.

—Naoaki Takashima

Nob Yoshigahara, our great puzzle friend, has passed away. I, like many of us, believed that Nob was indestructible. Nob witnessed the Hiroshima flash and aftermath as a child, he survived an explosion at his lab where he was teaching chemistry, he survived a kidnapping and mugging in New

York (it took Nob a long time to convince his friends that this was not one of his practical jokes), he survived stomach cancer, and more recently, he survived pneumonia. It is hard to believe that the world's most prolific mechanical puzzle designer and popularizer is no longer with us.

Nob has been my friend, mentor, inspirator, and puzzle father for a long time. Nob was a critical reviewer, helping me find good themes for puzzles and tune puzzles to an appropriate difficulty level. My first direct collaboration with him was on a key-shaped puzzle by Hanayama. Nob brought me in contact with (the late) Tadao Muroi, who made a perfect wooden prototype, that could be used by Hanayama. Nob ensured that I would get full credit by naming the puzzle after me. Nob helped select the best prototype for O'Gear and added my initial to the puzzle's name. The best example of our collaboration is most likely the Hanayama Chain puzzle. Nob suggested that the original design (later re-created by James Dalgety as "GGG") was far too difficult for the general public. So, I simplified the design as far as I could. Nob recruited Koji Kitajima to perfect the prototype. When I received the first samples, I could not even make the first move, and I complained to Nob. Nob just told me to try again. Only then did I recognize the extreme tolerance accuracy of the final puzzle. For over a year, we worked together on L'Oeuf, inspired by an idea by George Miller. Under Nob's supervision and using George's prototyping talent, the puzzle became more rugged, more complicated, and well-themed. It was Nob who came up with the perfect name for the puzzle. As recent as a few weeks before his death, Nob and I were communicating about new puzzle designs and ways to improve them. The photo here shows Nob and me during a recent visit to my house in Leidschendam, brainstorming how to get the best out of a puzzle idea.

Nob and van Deventer discussing puzzles at van Deventer's home.

Nob made a permanent mark on the international world of mechanical and mathematical puzzles. He stimulated a whole generation of puzzle designers, makers, collectors, and popularizers. Nob, we shall all miss you dearly.

—*M. Oskar van Deventer*

I've been privileged to know Nob, and to collaborate with him, for the past ten years. As head of Binary Arts (now ThinkFun), I've worked with Nob to bring six extraordinary and highly successful puzzles to market, as well as a good selection of extra challenges and extension products for the most popular of these puzzles.

I'd met Nob several times before, but he first visited our offices in 1995, along with Harry Nelson, who represented him in the United States. As a result of that visit, we started working together on Rush Hour, then called the Tokyo Parking Lot puzzle. It took nearly a year to figure out what to do with it, but in the fall of 1996, we introduced it to the market as Rush Hour. Since that time, we have sold nearly three million units of Rush Hour, and since its introduction, it has been our top selling product every year except one.

In addition to his creative genius, Nob had a genuine flair for marketing. Many of Nob's marketing ideas we didn't take to heart, which is perhaps our loss. With Rush Hour, Nob wanted us to organize New York taxi drivers and create a huge media event to launch the product. I've heard it rumored, though I don't know for sure, that he actually did organize such an event in Tokyo. Interestingly, we were contacted by MTV in 2001, and in the Spring of 2002 they devoted a whole program of their *Real World vs. Road Rules* show to contestants competing in a life sized version of Rush Hour, played with Saturn cars on the beach in Cabo San Lucas in Mexico. Later, once Nob was working with Hanayama in Japan in a big way, they produced a video loop of Nob pitching their line of disentanglement puzzles, which played in toy and game stores across the country and was very successful. Several times he strongly encouraged me to do the same in America. "One thousand Nobs, working around the clock, all across Japan, selling these puzzles," he told me. "One thousand Nobs!"

Those of you who knew Nob know that some ten years ago he battled stomach cancer and won, but lost his stomach in the process. Since that time he'd been able to eat little else but pureed foods. I suspect that he was in chronic pain for the last years of his life, though he continued to work through this, puzzling and figuring and inventing through all hours of the night. One time several years ago, Nob developed a severe infection. The medication that the doctors gave him for this had the effect of dulling his brain so that he couldn't think. For him, it was an extremely frustrating time. For me, it brought home the fact that Nob was an amazing person, really a force of nature. Earlier this year, when Nob was hospitalized for pneumonia and being fed intravenously, he insisted that only veins in his left arm be used, so that he could continue to type on his computer with the right arm.

As I reflect on Nob and my relationship with him, I am of course in awe of his abilities and deeply appreciative that I have been able to work with him as I have. But more than this, I'm appreciative of his humanity. For a tiny little man who seemed to run practically on fumes, Nob lived an outsized, even outrageous life. He drank, he smoked, he caroused. He was kind and sympathetic, he had friends all over the world, and he traveled extensively. There were times when you couldn't tell whether he'd be laid out by jet lag, his intestinal problems, drinking, or all of them simultaneously; then, he'd rally and his brain would be whirring again. And of course, he was very, very funny, in a way that only Nob could be. "Bad boy!" will stay with me, and with many of us, forever.

Bad boy, Nob! Rest in peace.

—Bill Ritchie

Part I

Braintreasures

Chinese Ceramic Puzzle Vessels

Norman L. Sandfield

I'm a clever teapot, yes it's true.
Here's an example of what I can do.
I can change my handle to my spout.
Just tip me over and pour me out!
— *I'm A Little Teapot*

Puzzle vessels are a popular curiosity around the world. They have been found in at least 24 countries and five or more ancient cultures, including those of Canaan, Greece, Phoenicia, Egypt, and Turkey. John Rausch (creator of the Puzzle World web site) calls them "the oldest known mechanical puzzles." While the author knows of over 30 museums in the world that own puzzle vessels, there are only a few with Chinese puzzle vessels, and none have more than a few examples.

Puzzle vessels are most often made of ceramic, although other materials, such as pewter, brass, and cloisonné have also been used. In China, many types of ceramics are used, including Celadon (green glazed stoneware, an attempt to duplicate opaque jade), Cizhou (Tz'u Chou, with darker brown details on a cream-colored background), Chinese blue and white, Rose Medallion (with decorated panels around a central medallion that includes a bird or peony), Yixing (the properties of the clay make it perfect for brewing tea), and Blanc de Chine.

Norman Sandfield is a long time dealer and author in the field of Japanese netsuke, and a collector in many diverse fields, who now owns almost 100 Chinese Ceramic Puzzle Vessels.

Many Chinese puzzle vessels look like—and some are—what we now call teapots. We use the umbrella term "vessel" to include the original Chinese vessel (which was probably a wine pot), the later teapot, as well as other puzzle pots and cups discussed later. It was not until the Ming Dynasty (1368–1644) in China that the teapot as we know it was found in common use. Prior to that, tea was made in and drunk from cups. Thus it appears that the earlier vessels from China may most accurately be called wine pots, while the English and other European puzzle vessels should be called teapots, even though many today are made strictly for puzzle use and purchased by tourists, rather than for actual drinking purposes.

While not directly related to the more familiar types of puzzles that are currently popular, such as crossword puzzles, jigsaw puzzles, and mechanical puzzles such as the famous Rubik's Cube, these vessels each contain a puzzle that must be solved. The puzzles are usually based on historically known laws of physics, such as the use of siphons. (See, for example, Woodcroft's translation [5] of "The Pneumatics of Hero, from the original Greek.") There are significant differences between Eastern and Western puzzle vessels, not just in appearance, but also in the goal of the puzzle. Western puzzle vessels commonly involve a person trying to drink from a mug, jug, or stein without spilling the liquid all over himself. The purpose of the puzzle is mainly the entertainment of others, often in a bar setting, when an unsuspecting patron is given the jug with an alcoholic beverage in it.

Eastern puzzles, including Chinese puzzle vessels, are more often about *filling* the vessel through a non-obvious hole, often in the bottom of the vessel. An example is the *Dao Liu Hu*, or Reversed Flowing Pot. The one shown in Figure 1, made at the Yaozhou Kiln factory in Xi'an, is an imitation of Yaozhou porcelain of the Song Dynasty (960–1279 C.E.). It has a spherical shape, and the outside surfaces are cut with many beautiful flower patterns, indicating luck. The handle of the pot is a flying phoenix, and the mouth is the mouth of a small lion that is being suckled. There is a small hole in the center of the foot of the pot. Water is poured in through the hole, and then the pot turned back upright. Because of an internal tube, the pot doesn't drip from the hole; if you want to pour the water out, it will run out from the spout, as a normal pot.

The Chinese puzzle cup is known variously as the Justice Cup, the Fairness Cup, and the Greedy Cup. (Western versions are sometimes called Tantalus Cups, after the mythological Greek figure Tantalus, whose punishment for annoying the gods was to stand neck-deep in water that disappeared whenever he tried to drink.) Called a *Gong Dao Bei* in Chinese, it traditionally, though not always, has a matching base to hold the spilled water that drains out. The ones shown in Figures 2 and 3 have a standing

Figure 1. Reversed Flowing Pot. Late 20th century.

Figure 2. Justice Cup with base. Late 19th / early 20th century. (See Color Plate III.)

dragon head coming up from the bottom of the cup and a small hole in the foot of the cup. Rather than being used to embarrass guests and amuse onlookers, the Justice Cup is intended to illustrate a lesson of moderation. If you fill the cup to below the brim, the water will stay in the cup. But if you fill the cup all the way to the top of the brim, a siphon effect is created, and all of the water will drain out the bottom. The manufacturer of a *Gong Dao Bei* similar to in Figure 2 includes a story [7]:

> According to legend the cup is a family heirloom of a Tang Emperor. At the wedding ceremony Li Mao (who is the prince of the emperor Li Longji) and Yang Yuhuan, Li Longji gave the cup as a present to Yang Yuhuan and asked her: "Do you know the meaning why I give the cup to you?" Yang answered: "Doing anything should be controlled moderately, or nothing will be accomplished." [adapted from [7]]

There are many other kinds of Chinese puzzle vessels, of which there are too few of each type known to make a separate category for each. The Floating Man Cup (Figure 4) has a small porcelain figure inside an inverted dome in the base of the cup, which rises with the level of liquid in the cup.

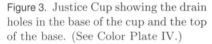

Figure 3. Justice Cup showing the drain holes in the base of the cup and the top of the base. (See Color Plate IV.)

Figure 4. Floating Man Cup from the late 19th or early 20th century. Polychrome, $1\frac{3}{4}$ inches high. From the collection of Dottie and Bob Wallace.

Double Pour Pots have two separate compartments, often solved by an appropriate placement of thumb or fingers over various holes—the ideal teapot to poison someone! There are gourd-shaped wine pots where the spout is also the handle, rotate-to-pour teapots, three-piece bottom-pour teapots, top-fill and top-pour pots, current Chinese copies of European-style puzzle steins, and vessels with hidden lithophane images. (A lithophane is a thin piece porcelain with a picture carved in it, which can be seen when held to the light.) Some of these are definitely teapots, rather than wine pots, for two reasons. First, they appear to be made of Yixing ware, which is the preferred material for teapots in China. Second, they all have a built-in strainer in the base, above the pour hole, to strain out the tea leaves.

It could be argued that another category of Chinese ceramic puzzle vessels, currently found throughout China and in a variety of models (mostly Yixing ware, and some identical models in a white porcelain, in particular), is the "pissing man" or "pissing pig." Prior to presentation, these are filled with water. Then, when hot water is poured over them, air pressure forces the water inside to come out through a very small upward-pointing hole, making the human or animal look as if he is urinating. These are fun objects, but not really puzzles—the most puzzling aspect of them is how to get the water inside the sealed ceramic object in the first place.

Figure 5. Bottom-fill cup from the Song Dynasty (960–1279 C.E.). Probably the oldest puzzle vessel in the author's collection, and one of the oldest puzzles in the world.

Figure 6. Bottom-fill cup from the late 19th or early 20th century. Peachshape, 6 inches high, Celadon ware. From the collection of Dottie and Bob Wallace.

Figure 7. Blanc de Chine cup, from the late 18th to early 19th century. 2 inches high.

Figure 8. One of about ten peachshaped, bottom-fill wine pots recovered in the 1980s from a Chinese boat that sank in around 1643. Put up for sale in June, 1999, at Ben Janssens Oriental Art Limited, London, for 19,000 pounds (approximately $30,000).

Acknowledgments

This article grew out of a presentation at IPP 19 in London, 1999, and a monograph presented at G4G4, Atlanta, 2000. The author would like to thank Jerry Slocum, Harold Raizer, Les and Marci Barton, Robert Koeppel, Dottie and Bob Wallace, Chase Gilmore, James Thornton, Dr. Bennet Bronson, Frans de Vreugd, David Singmaster, Wei Zhang, and Peter Rasmussen for their significant help. Except for the wine pot in Figure 8, the vessels illustrated are currently in the author's collection. The photographs are by David A. Weinstein and Associates Ltd., Chicago, and used with permission.

References

[1] de Vreugd, Frans, "Oriental Puzzle Vessels." *Cubism for Fun (CFF)*. 1999, The Netherlands: Dutch Cubist Club (DCC). Pages 18–20, 12 puzzle vessels in color and 1 illustration.

[2] de Vreugd, Frans, "Chinese Puzzlehunt." *Cubism for Fun (CFF)*. 2001 (56, October) The Netherlands: Dutch Cubist Club (DCC): Pages 20–24, 8 color illustrations.

[3] Gorer, Edgar and James F. Blacker, *Chinese Porcelain and Hard Stones.* Illustrated by Two Hundred and Fifty-Four Pages of Gems of Chinese Ceramic and Glyptic Art. vol. 1. 1911, London England: Bernard Quaritch. Plate 135 illustrates two very similar, yet completely distinctive, types of Chinese puzzle vessel, more closely related in puzzle technique to English puzzle jugs or steins. These are how-to-pour or drink puzzle vessels.

[4] Sandfield, Norman L., *A Monograph on Chinese Ceramic Puzzle Vessels (Antique, Vintage and Contemporary).* 1999, 2000, Chicago, Illinois. 37 pages; with color illustrations on the cover and the back. Limited edition of less than 150 copies.

[5] Slocum, Jerry [Gerald K.] and Jack Botermans, *Puzzles Old and New: How to Make and Solve Them.* 1986, 1987: Amsterdam, The Netherlands: Plenary Publications International, 1986; Seattle: University of Washington Press, 1987. Three Chinese wine pots illustrated in the middle of page 140; two are peach shaped and one has a large Foo Lion Dog on it. Gives an outline of the ancient history.

[6] Woodcroft, Bennet, translator and editor, *The [Treatise on] Pneumatics of Hero*, from the original Greek. 1851, London, England: Taylor Walton and Maberly. Available on the Internet at http://www.history.rochester.edu/ steam/hero/. Includes 78 fascinating mechanical objects using often complex principles of pneumatics to create, what would have been at that time, magical effects and sections on siphons related to the functioning of the Greedy cup.

[7] Yaozhou(ware) Kiln Museum. 1998: Shanghai, China. Several brochures about the history of the factory and its products, only one of which mentions the Celadon Reversed Flowing Pot. Plus 2 descriptive pamphlets, packaged with each of the Celadon puzzle vessels, when sold in the nicer brocade boxes. Chinese and English text.

Mongolian Interlocking Puzzles

Jerry Slocum and Frans de Vreugd

In August 2000, one of us (Jerry) was visited by an antique dealer who spends most of his time searching for rare antiques in places that most of us collectors would consider too remote or too dangerous. The dealer enthusiastically described an amazing Puzzle Museum that he had just found in a most unlikely place, Ulaanbaatar, Mongolia. "There are more than 2,000 puzzles; you must go there," he said. He provided the business card of the Director, Mr. Zandraa Tumen-Ulzii, drew a map of the location of the museum, and mentioned that Mr. Tumen-Ulzii's daughter, Monica, was living in Los Angeles!

Ulaanbaatar, the capital of Mongolia, is located between northern China and Russia's Siberia. The country is twice the size of Texas, and the average altitude is about 5,000 feet. It has very cold winters, and even during the summer it is frequently quite cool. Historically Mongolia has been a nomadic society. Even today many people live in round, easily moved,

---

Jerry Slocum is the author of nine books about mechanical puzzles and is also known for his research on the history of puzzles and his large collection of puzzles and puzzle books. He is the founder and organizer of the annual International Puzzle Parties, held in the United States, Europe and Asia.

Frans de Vreugd is a Dutch puzzle designer and collector, traveling around the world in the search for puzzles. He is one of the editors of the puzzle newsletter *Cubism For Fun*.

Figure 1. Mr. Tumen-Ulzii.

felt-covered *Gers* (tents) while tending herds of horses, sheep, yaks, goats, and camels. There is little agriculture and limited industry in the country. With the help of Monica and her brother Itgel, a trip to Mongolia was arranged. Jerry invited puzzle friends Dick Hess and Frans de Vreugd to accompany him on the trip. Our adventure in Mongolia is described in reference [1].

Mr. Tumen-Ulzii, Puzzle Inventor, Artist, and Craftsman

Mr. Tumen-Ulzii was born in 1944 in Aguit, Mongolia. He was one of eleven children. His parents were herdsmen. He is now married and has two sons and two daughters. He and all of his children are University graduates.

Mr. Tumen-Ulzii became interested in puzzles at a very young age when he was given a six-piece burr puzzle by his father with one of the pieces missing. He soon figured out the design of the missing piece, carved a replacement and solved the puzzle. He was hooked on puzzles!

Since 1955, Mr. Tumen-Ulzii has invented, patented and crafted more than 2,445 different interlocking mechanical puzzles. He opened his International Intellectual and Puzzle Museum in Ulaanbaatar in 1990 in order to share his puzzles and chess sets with other Mongolians and visitors from all over the orld.

The work of Mr. Tumen-Ulzii can best be described as a delicate combination of intricate puzzle design, art, and very skillful craftsmanship. He has used these talents to design and fabricate a completely new class of

Figure 2. Puzzle art by Tumen-Ulzii. (See Color Plate IX.)

interlocking puzzles, unsurpassed in its kind. Mr. Tumen-Ulzii has designed puzzles that range from simple (but beautifully decorated) ones to extremely complicated interlocking structures using hundreds of pieces. His puzzles use many different techniques, including dovetails, rotational, and tilted moves.

Mr. Tumen-Ulzii has been recognized for his many accomplishments. He was awarded a prize as the "Best Inventor of Mongolia" in 1994 by the Mongolian Government. And in 1998, the President of Mongolia awarded him Mongolia's highest honor, a gold medal and the title of "Meritorious Person of Culture of Mongolia," "for his contribution to developing the minds and thinking skills of children and youth."

Puzzle Art and Jewelry

In addition to designing and making puzzles, Mr. Tumen-Ulzii has painted several pictures which feature puzzles with symbolic meaning. The picture on the left in Figure 2 shows a baby's pacifier in the shape of a puzzle, meaning that we should not only nourish our young children with food, but also feed them intellectually. Mr. Tumen-Ulzii has also designed and made a unique four-band puzzle ring, shown in Figure 3.

Figure 3. Puzzle ring.

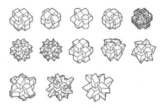

Figure 4. Six-piece burr variations.

Six-Piece Burrs

Although most of the six-piece burrs Mr. Tumen-Ulzii has designed are
relatively simple as puzzles, others require as many as seven moves to
remove the first piece and seventeen moves to disassemble completely. All
of them are extremely decorative. He has come up with several dozen
different ways to decorate the pieces. The ends of the pieces of some of
the puzzles have an unusual shape, shown in Figure 4. Other puzzles
are miniature sculptures with intricate hand carved decorations. Figure
5 shows one with the ends of the pieces carved and decorated as Donald
Duck and Mickey Mouse and another with the twelve traditional animals
of the Chinese Zodiac carved and painted on the pieces.

Figure 5. Disney characters and Zodiac figures decorate the six-piece burrs.

Figure 6. Decorated 12- and 27-piece burrs.

More Complex Burrs

Mr. Tumen-Ulzii moved up to a higher level of complexity with the 12- and 27-piece burrs shown in Figure 6. And each piece in the 18-piece Board Burr shown in Figure 7 is different and quite complex, making this one of the most difficult puzzles we found in the museum. The orientation of the pieces is utterly confusing. Not only do the ends alternate, but there are different arrangements in different directions (2×3 in the x and y directions and 1×6 in the z direction). Many of his interlocking puzzles contain many more pieces. The puzzle shown in Figure 8 contains 375 pieces. Another puzzle he designed, called "Cosmic Eden," uses 673 wooden pieces.

Figure 7. 18-piece Board Burr and one of its pieces.

Figure 8. 375-piece puzzle.

Figural and Decorative Interlocking Puzzles

Several of Mr. Tumen-Ulzii's more complicated puzzles, such as those shown in Figure 9, are decorated by beautifully carved animal shapes.

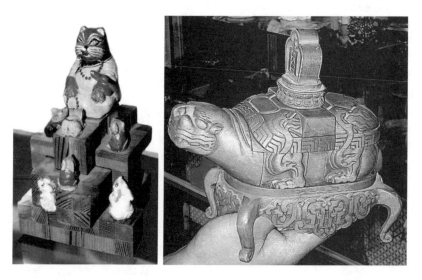

Figure 9. Carved and decorated wooden Cat and Mouse Puzzle and Turtle Puzzle. (See Color Plate VIII.)

Figure 10. Zodiac figural puzzles. (See Color Plate II.)

As with previous puzzles, the decorations frequently have a traditional symbolic meaning. Figure 10, for example, shows four out of a complete set of twelve animals representing the signs of the Zodiac. The internal structure and mechanisms to disassemble and reassemble the puzzles are all different and vary in difficulty. From the outside it is often unclear how many pieces the puzzle has.

Figure 11 shows one of his best puzzle designs. It is a turtle-shaped interlocking puzzle, made from eleven pounds of silver! This turtle is also the symbol of the city of Ulaanbaatar. It requires nine complex moves to remove the first piece and 33 moves to completely disassemble it. Mr. Tumen-Ulzii has offered $100,000 to any person who can take it apart and

Figure 11. Silver Turtle Puzzle.

reassemble the puzzle in ten minutes. We tried, but could not even get the first piece out in ten minutes. Some of the movements required are very unusual. We doubt that even an experienced solver can solve the puzzle in the given time.

Interlocking Buildings

Mr. Tumen-Ulzii has also designed several impressive interlocking puzzles in the shape of famous buildings. Figure 12 shows a fairly simple seven-piece Statue of Liberty and a puzzle in the shape of a traditional Mongolian *Ger* (tent). Other impressive puzzle structures in the museum include the Eiffel tower as well as the Chinggis Khaan Hotel and the statue of Sükhbaatar in Ulaanbaatar (both very difficult).

Figure 12. Statue of Liberty and Mongolian *Ger* puzzles.

Puzzle Chess Sets

One of Mr. Tumen-Ulzii's passions is playing chess. He has designed more than 20 different puzzle chess sets. In most of these, each type of piece (six types of each color) are distinct puzzles, ranging from easy (pawns) to difficult (king/queen). Not only are all the chess pieces puzzles, but so are the joints in the accompanying table. There is a great diversity in size, shape and material. For example the smallest chess set is about 20×20

Figure 13. Turtle chess set. (See Color Plate X.)

cm, and the pieces are tiny. Another chess set, which was only finished the day before we arrived, uses animal figures for all of the pieces, which become decorative as well as intriguing puzzles. The chess table, shaped like a big turtle, is shown in Figure 13. The pawn, the simplest puzzle, uses a 180 degree rotation as first move. Mr. Tumen-Ulzii's most precious chess set, shown in Figure 14, uses gold for the black chess pieces and silver for the white pieces. The intricate design of the six pieces of the Chess King indicates the difficulty of disassembling and assembling the puzzle. Eight moves are needed to remove the first piece and eighteen moves to disassemble it completely. The Chessboard includes 108 jewels and precious stones. In May 2002, as part of the 840th anniversary celebration of Chinggis Khaan, he made the world's largest Mongolian Chess set with pieces two feet high and a chessboard 25 feet square. In this chess set the two kings and the board are puzzles.

Figure 14. Gold and Silver Chess Set with six pieces of the King on the right.

Conclusion

The Mongolian Puzzle Museum and Mr. Tumen-Ulzii are amazing. The richness and diversity of interlocking puzzles in the museum is unmatched anywhere in the world. Mr. Tumen-Ulzii has combined his inventiveness and technical design skills with his unique talent as an accomplished artist and sculptor for the design and fabrication of thousands of complex interlocking puzzles. His museum has since moved to a much larger building. We plan to visit his new museum and bring many other puzzle enthusiasts with us.

References

[1] Frans de Vreugd. "Puzzle Adventures in Mongolia." *Cubism for Fun 57*, Dutch Cubists Club (NKC), March 2002.

Part II

Brainticklers

Fold-and-Cut Magic

Erik D. Demaine and Martin L. Demaine

Fold-and-Cut Problem

Take a sheet of paper, fold it flat however you like, and make one complete straight cut. What are the possible shapes of the unfolded pieces? For example, Figure 1 shows how to produce a five-pointed star by folding and one straight cut. You could imagine cutting out the silhouette of your favorite animal, object, or geometric shape.

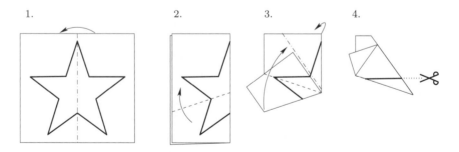

Figure 1. How to fold a square of paper so that one cut makes a five-pointed star.

The Demaines are a father-and-son team. **Erik D. Demaine**, the son, is an assistant professor of computer science at MIT who likes anything algorithmic or geometric. **Martin L. Demaine**, the father, is a researcher in computer science at MIT who likes anything mathematical and artistic.

Martin Gardner wrote about this *fold-and-cut* problem in a 1960 article in *Scientific American* called "Paper Cutting" [7]. In addition to providing some of the historical references below, he was the first to pose the then-open problem: What are the limits of this fold-and-cut process? What polygonal shapes can be cut out?

History

The first published reference we know of is a Japanese puzzle book [12] by Kan Chu Sen in 1721. As one of many problems testing mathematical intelligence, this book asks the reader whether it is possible to fold a rectangular piece of paper flat and make one complete straight cut so as to produce a typical Japanese crest called *sangaibisi* (Figure 2), which translates to "three folded rhombics." Towards the end, the book shows a solution.

Another early reference to folding and cutting is an 1873 article in *Harper's New Monthly Magazine* [1]. This article may be the first written account of a wide-spread tale of how Betsy Ross came to sew the first American flag. The story goes that, in 1777, George Washington and a committee of the Congress showed Betsy Ross plans for a flag with thirteen six-pointed stars, and asked her whether she could make such a flag. She said that she would be willing to try, but suggested that the stars should have five points. To support her argument, she

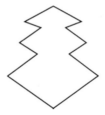

Figure 2. Japanese crest from [12].

demonstrated how easily a regular five-pointed star could be made by folding a sheet of paper flat and making one cut with scissors. The committee accepted her changes, and George Washington made a new drawing, which Betsy Ross followed to make the first American flag.

Folding and cutting has naturally also had its place in the magic community, and it is with these references that Martin Gardner was primarily familiar. Will Blyth was perhaps the first with his 1920 book *Paper Magic* [3], which includes a Maltese cross, a six-pointed star, and an intricate multipiece altar. Harry Houdini was a general magician before he became the famous escape artist, and his 1922 book *Paper Magic* [8] describes a method for making a five-pointed star, similar to the one in Figure 1. This book may have in fact been ghost-written by fellow magician Walter Gibson [9]. Another magician, Gerald Loe, studied the fold-and-cut problem in some detail; his 1955 book *Paper Capers* [10] describes how to cut out arrangements of various geometric objects, such as a circular chain of stars. Gardner was particularly impressed with Loe's ability to cut out most letters of the alphabet. Gardner surveys a variety of other one-cut tricks in his *Encyclopedia of Impromptu Magic* [6].

Mathematics

Around the summer of 1997, Anna Lubiw and the present authors set out to answer Gardner's question from a computational geometry perspective: given a target shape, is there a good algorithm to determine whether it can be produced, and if so, to design a flat folding and say where to cut?

The surprising answer is that *every* polygonal shape can be produced. In fact, any desired pattern of cuts can be simultaneously produced, by folding and one straight cut. Such patterns allow us to make several shapes at once, to make holes within shapes, and to dissect the piece of paper into multiple adjoining shapes. Furthermore, there are algorithms that tell us where to fold and where to cut.

One solution and algorithm for the problem is by the original investigators [4, 5]. This solution is based on a structure called the "straight skeleton" which captures symmetries in the target pattern of cuts. Later on, Marshall Bern, David Eppstein, and Barry Hayes brought a new disk-packing approach to the problem, resulting in a second solution and algorithm for the problem [2].

These algorithms allow us to design many practical examples of the fold-and-cut idea that go far beyond what was previously possible. The first solution [4, 5] produces more practical foldings, so our examples in the next section use this algorithm.

Before we turn to examples, however, we should mention one minor catch in applying the general mathematical theory to practice, namely the precise definition of a "cut." In one model, cuts can be made along creases, effectively slitting the paper; in the other model, cuts cannot be made along creases and must be slightly separated. The first model is more natural mathematically, but the second model is more practical.

The general result stated above applies only to the first model of cuts. For the second model, the patterns that can be produced by a single cut are precisely those where an even number of line segments meet at any point. All examples described in this article have this evenness property, and therefore obey the second, more practical model of cutting. More generally, this issue arises only for dissections; cutting out a single shape or even several disjoint shapes is not a problem because precisely two line segments meet at any point.

Examples

Figures 3, 4, 7, 8, and 9 show several examples of how to produce shapes by one straight cut.[1] All examples are based on the original algorithm [4, 5]. The examples are designed to exploit symmetry and alignment to make them as easy as possible to fold. Naturally, some are easier than others.

The figures use the following notation. The shapes to be cut out are in thick lines; these lines are not creased. Dashed lines denote *valley* folds, where the paper is folded "towards" you. Dot-dashed lines denote *mountain* folds, where the paper is folded "away" from you. Some of the figures have a line of symmetry, in which case you must first fold along that line of symmetry, and then apply the remaining creases. In general, however, all creases must be folded more or less simultaneously.

The first example (Figure 3) produces the initials of the *Gathering for Gardner 5*. It also has five pieces, counting the exterior silhouette.

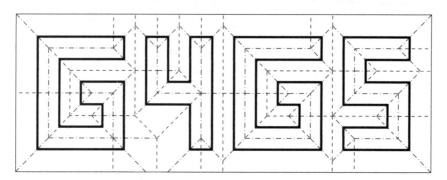

Figure 3. Initials of the Gathering for Gardner 5.

The second example (Figure 4) produces a cross shape along with the letters H-E-L-L, again consisting of five pieces. This example is a redesign of a classic fold-and-cut magic trick described by Gardner [6, 7] and shown in Figure 6. The classic trick is easier to fold, but requires the letters to be formed out of several smaller pieces. As a challenge, we considered what would be possible if we used an entire rectangle of paper to cut out precisely the four letters and the cross. It is difficult to design a version with good proportions between the letters. For example, the more proportionate dissection in Figure 5 has an odd number of line segments coming together at some points, so it is unsuitable unless you allow cutting along the creases.

---

[1] All foldings (except the classic in Figure 6) were designed by the present authors. Figures 7 and 9 have appeared before [4, 5].

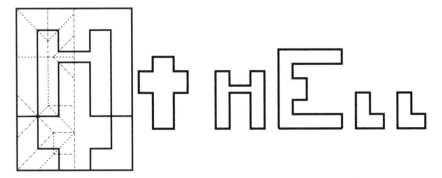

Figure 4. Cross and H-E-L-L. Fold in half first, then use the specified creases.

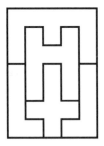

Figure 5. Unsuitable dissection.

For those unfamiliar with the cross/hell magic trick, the patter goes something like this. Two people, Good and Evil, arrive at the gates of heaven, but only Good has a ticket. Evil begs Good for help, so Good folds his ticket as shown, rips along a line, and hands Evil the smaller pieces. Unsure of what to do with the pieces, Evil hands them to St. Peter, who re-arranges them to spell H-E-L-L, to which Evil is appropriately directed. Good hands the remaining piece to St. Peter, who is pleased to unfold a cross.

Figure 6. Classic method for producing cross and multipiece H-E-L-L.

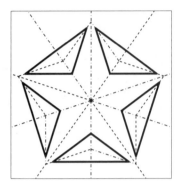

Figure 7. 5 triangles in a star pattern.

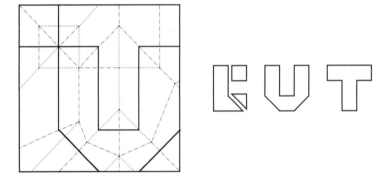

Figure 8. 5-piece dissection of C-U-T.

The third example (Figure 7) consists of five triangles arranged in the pattern of a five-pointed star [4, 5]. Gerald Loe [10] designed a similar example.

The fourth example (Figure 8) is a five-piece dissection of the square whose pieces can be re-arranged to spell the word CUT. The underlying dissection was designed by Joseph Madachy [11].

The next three examples (Figure 9) are silhouettes of three animals: a swan, butterfly, and angelfish [4, 5].

Acknowledgment

We thank Martin Gardner for inspiring us to work on this problem and thereby enter the area of computational origami.

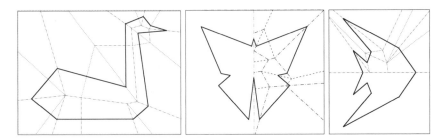

Figure 9. Swan, butterfly, and angelfish.

References

[1] National standards and emblems. *Harper's New Monthly Magazine*, 47(278):171–181, July 1873.

[2] Marshall Bern, Erik Demaine, David Eppstein, and Barry Hayes. A disk-packing algorithm for an origami magic trick. In *Proceedings of the 3rd International Meeting of Origami Science, Math, and Education*, pages 17–28, Monterey, California, March 2001. Improvement of version appearing in *Proceedings of the International Conference on Fun with Algorithms*, Isola d'Elba, Italy, June 1998, pages 32–42.

[3] Will Blyth. Gains of the great war. In *Paper Magic: Tricks and Amusements with a Sheet of Paper*, pages 95–99. C. Arthur Pearson, Limited, London, 1920.

[4] Erik D. Demaine, Martin L. Demaine, and Anna Lubiw. Folding and cutting paper. In J. Akiyama, M. Kano, and M. Urabe, editors, *Revised Papers from the Japan Conference on Discrete and Computational Geometry*, volume 1763 of *Lecture Notes in Computer Science*, pages 104–117, Tokyo, Japan, December 1998.

[5] Erik D. Demaine, Martin L. Demaine, and Anna Lubiw. Folding and one straight cut suffice. In *Proceedings of the 10th Annual ACM-SIAM Symposium on Discrete Algorithms*, pages 891–892, ACM Press, Baltimore, MD, January 1999.

[6] Martin Gardner. Single-cut stunts. In *Encyclopedia of Impromptu Magic*, pages 424–428. Magic, Inc., Chicago, 1978.

[7] Martin Gardner. Paper cutting. In *New Mathematical Diversions (Revised Edition)*, chapter 5, pages 58–69. The Mathematical Association of America, Washington, D.C., 1995. Appeared in *Scientific American*, June 1960.

[8] Harry Houdini. *Paper Magic*, pages 176–177. E. P. Dutton & Company, New York, 1922. Reprinted by Magico Magazine.

[9] G. Legman. *Bibliography of Paper-Folding*. Priory Press, Malvern, England, 1952.

[10] Gerald M. Loe. *Paper Capers*. Magic, Inc., Chicago, 1955.

[11] Joseph S. Madachy. Geometric dissections. In *Madachy's Mathematical Recreations*, chapter 1, pages 15–33. Dover Publications, New York, 1979. Reprint of *Mathematics on Vacation*, Scribner, New York, 1975.

[12] Kan Chu Sen.   *Wakoku Chiyekurabe (Mathematical Contests)*.   1721. Excerpts   available   from   http://theory.lcs.mit.edu/ẽdemaine/foldcut /sen_book.html.

The Three-Legged Hourglass

M. Oskar van Deventer

Every time my parents come over, my wife prepares a big breakfast with boiled eggs. Boiling the eggs right is difficult. My father likes his egg boiled 4 minutes, my mother 2 minutes, my wife 7 minutes and I prefer my egg boiled 3.5 minutes. I would need to have four hourglasses for all these wishes, or more when I get other guests.

Figure 1. Three-legged hourglass.

My solution is the Three-legged Hourglass, as shown in Figure 1. This hourglass contains 8 minutes worth of sand altogether. When it is turned,

M. Oskar van Deventer is the creator of hundreds of innovative mechanical puzzle designs, several of which are commercially available.

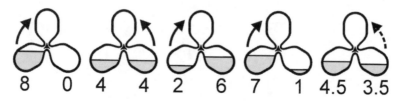

8   0   4   4   2   6   7   1   4.5   3.5

Figure 2. Operating instruction.

the sand in the top compartment equally divides over the two other compartments. Figure 2 shows the operation sequence that takes care of my family. First collect all 8 minutes of sand in one compartment by holding the hourglass with one point down. Then turn over the hourglass and wait till all sand has run through. Then turn the glass left and the compartment empties in 4 minutes. After that turn it right and it empties in 2 minutes. Once more and it empties in 7 minutes. Finally, turn it left and it empties in 3.5 minutes. And breakfast can begin!

The Incredible Swimmer Puzzle

Stewart Coffin

This amazing story involves the unlikely combination of the puzzle page in the Old Farmer's Almanac, the pitching coach of the Cleveland Indians, quadratic equations, the Reference Librarian of the Amesbury Public Library, and finally, if you can believe it, rivers flowing upstream.

It all begins back in 1969, when I started handcrafting unusual wooden puzzles. At the same time, I decided to try inventing some mathematical puzzles and perhaps getting them published.

After submitting about half a dozen puzzles for possible inclusion in the puzzle page of the Old Farmer's Almanac, one of them was finally accepted for publication. I soon became so involved with woodworking that I thought no more about math puzzles, until about a year later when I started receiving letters about my submission. It seems that no one could figure out how to get the published answer. To my chagrin, I then discovered that one essential fact had been carelessly omitted from the statement of the puzzle, making it impossible to solve. I had not saved a copy of my submission, so to this day we don't know whose mistake it was, but it was probably mine.

Stewart Coffin is recognized as the world's best designer of polyhedral interlocking puzzles. He is the creator of AP-ART and author of several books on puzzle craft.

By far the most interesting letter received was from Cot Deal, pitching coach of the Cleveland Indians. Evidently my puzzle had precipitated an argument among him and some of the players, and they all wanted to know the explanation for the answer that was published. This resulted in a friendly exchange of letters, in one of which he invited me and my family to be his guests the next time the Indians were at Fenway Park. We immediately accepted his invitation.

After the game, we waited around where the players came out in order to meet coach Deal and thank him. He said he didn't know anyone else in the Boston area and invited us to be his guests whenever the Indians were in town. Unfortunately, the Indians were having a bad season, manager Alvin Dark was soon replaced, and probably likewise coach Deal because we never heard from him again.

Now switch ahead thirty years. In my old age I have been occupying myself by writing down my recollections, mostly for family and friends. I had a notion to include this story pretty much as written above. While doing so I searched diligently for any related papers, such as the original version of my puzzle or the Deal correspondence and sadly came up with a complete blank. I assumed that they were probably lost years ago, along with so many other things, when I moved from Lincoln, MA to Andover, MA.

Then quite recently, while looking for something else, what should I stumble upon but the long lost Deal letter. That letter established, first of all, the year of the OFA article as being 1970, which I had forgotten. Second, it gave the published answer. In the same file was another lucky find—a scrap of paper with some calculations on it, which I recognized as my work sheet for solving the puzzle.

I then decided to attempt to reconstruct the original puzzle by working backwards from the solution and those few scribblings of calculations—not an easy task, especially for an old man approaching senility.

Next I turned my attention to locating a copy of the 1970 OFA. They are not so easy to find now. Finally, after an extensive search, what should come to my rescue but the Merrimack Valley Library Consortium. It seems that we can now search for library materials just about anywhere in eastern Massachusetts, thanks to a nifty computer database. Of all the libraries in our region, only one listed the long sought 1970 issue of OFA, and that one was the Amesbury Public Library. Reference Librarian Margie Shepard immediately offered to send me a photocopy of my long lost puzzle submission.

With all of this information finally at hand—the printed version of the puzzle, the published solution, and my original calculation notes—I was at long last able to reconstruct the whole thing. Here is the way it should have appeared:

A man rows upstream for one hour, and then for some strange reason jumps overboard and swims back downstream to his starting point. Meanwhile his boat drifts back down with the current and (this is the part that was accidentally omitted) arrives back at the starting point two hours after the swimmer. By the way, in still water the man can row twice as fast as he can swim. Question: How much time could he have saved by rowing back instead of swimming?

It is an unusual and rather confusing motion problem because no distances are given and hence no speed can be known, only time. Unless someone can come up with a better method, the only solution that I know involves the use of a quadratic equation. This yields two answers—the intended positive one and another that is negative.

I thought that might be the end of this story, but then I was always puzzled by that negative solution. Do you suppose that there might be some sort of bizarre interpretation, with time going backward or water flowing upstream?

Solution to the Incredible Swimmer Puzzle

Let r denote the speed of the river in miles per hour, and let s denote the man's swimming speed in still water. From these variables we can compute the man's upriver swimming speed $(s - r)$, the man's downriver swimming speed $(s + r)$, the man's still-water rowing speed $(2\ s)$, the man's upriver rowing speed $(2\ s - r)$, and the man's downriver rowing speed $(2\ s + r)$.

First the man rows upstream for one hour, i.e., $2\ s - r$ miles. Then the man swims downstream back the same distance, which takes $(2\ s - r)\ /\ (s + r)$ hours. It takes the boat two hours longer than that to travel the same distance at speed r. Thus

$$2 + (2s - r)/(s + r) = (2s - r)/r,$$
$$(4s + r)/(s + r) = (2s - r)/r,$$
$$r(4s + r) = (s + r)(2s - r),$$
$$4rs + r^2 = 2s^2 + rs - r^2,$$
$$2s^2 - 3rs - 2r^2 = 0.$$

This quadratic equation has two solutions: $s = 2\ r$ and $s = -r/2$. We only consider the positive solution of $s = 2\ r$, i.e., the man swims in still water twice as fast as the river flows. The man spent $(2\ s - r)\ /\ (s + r) = 1$ hour swimming back. If he had instead rowed the $2s - r$ miles, it would have taken him only $(2s - r) / (2s + r) = 3/5$ hours. Hence he would have saved $2/5$ hours $= 24$ minutes if he had rowed instead of swum.

The Butler University Game

Rebecca G. Wahl

The mathematics and actuarial science department at Butler University is capitalizing on the educational opportunities afforded by puzzles and games. Longtime Martin Gardner devotee and puzzle and games expert Jeremiah P. Farrell, professor emeritus, has provided the expertise and inspiration for students and faculty to enjoy learning new and easily accessible mathematics. Games are being used to entice prospective mathematics majors, to introduce mathematics majors to undergraduate research, and as a vehicle for an outreach program with the Indiana School for the Blind.

The department commissioned Prof. Farrell to create a customized "Butler University" game and puzzle to send prospective mathematics majors. The game is based on a simplified version of the Butler University campus map depicting key campus buildings shown in Figure 1. We imagine a snowstorm so severe that sidewalks consist of narrow passages that have been marked by grounds crews with only one direction of travel. Graph theorists call the underlying construction a directed graph or, more simply, a digraph. Several of Farrell's games are based on digraphs; once students are captivated by the games, the necessary mathematical theory becomes more relevant and interesting.

A two-person game and two puzzles are played on this digraph. In the two-person game, tokens (any number of coins will do) are placed on the nodes of the grid. Players alternately select a coin and move it along a

---

Rebecca G. Wahl is an assistant professor at Butler University in Indianapolis.

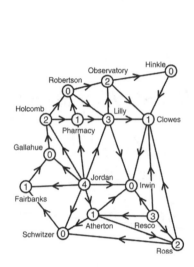

Figure 1. The directed graph on which the Butler University Game is played.

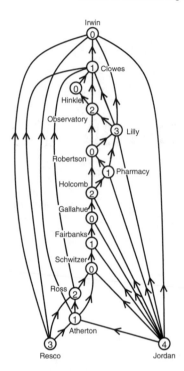

Figure 2. A redrawing of the directed graph from Figure 1 in which every arrow points upward.

directed edge to an adjacent node. Several coins may be stacked on the same node but only one coin may be moved during a player's turn. The game ends when all the coins are at Irwin Library, the winner being the last player to make a move.

It's not immediately obvious the game must eventually come to an end, but this becomes obvious when the digraph is redrawn as in Figure 2. In this drawing, all of the arrows point upwards, and therefore the graph has no cycles. (In fact, every directed acyclic graph has such a drawing [2].)

Readers familiar with the game Nim will be able to play this game expertly. For details we refer the reader to the article by Fraenkel [3], the book and article by Guy [4, 5], and the classic book *Winning Ways* [1]. In the puzzles, we offer a specific board configuration and request the winning move.

Puzzle 1: Suppose the board has exactly four coins on the same node. Who should win this game, the first or the second player?

Puzzle 2: Suppose one coin is placed at each node. It is your move. You have 28 moves to choose from, only one of which leads to a forced win. Can you find it?

Puzzle 1 has a simple solution: The second player should win the game, by matching the first player move for move, always restoring the configuration to an even number of coins on each node. Puzzle 2, though, calls for closer analysis.

Solutions to puzzles such as these have been studied by Fraenkel [1], where the winning strategy is obtained by calculating the Sprague-Grundy function of the underlying digraph. The values of the Sprague-Grundy function, called Nimbers, are written as the digits 0, 1, 2, 3, and 4 next to the nodes in Figure 1. They are computed by working downwards in Figure 2: Assign 0 to the highest node(s); to each lower node, assign the *smallest* value *not* assigned to any of the nodes it points to. For example, Pharmacy gets the value 1 since it points to Robertson (0) and Lilly (3).

Although Nimbers look like ordinary numbers, they are not. In particular, they use a different kind of addition. The "Nim-sum" of two Nimbers is obtained by binary addition without carrying. For example, 2 and 3, written in binary, are 10 and 11, hence their Nim-sum is 01, or simply 1. One easy consequence of this definition is that $a + a = 0$ for any Nimber a. We give a Nim-sum table for the first eight Nimbers below:

0	1	2	3	4	5	6	7
1	0	3	2	5	4	7	6
2	3	0	1	6	7	4	5
3	2	1	0	7	6	5	4
4	5	6	7	0	1	2	3
5	4	7	6	1	0	3	2
6	7	4	5	2	3	0	1
7	6	5	4	3	2	1	0

When coins are placed on the board, it is convenient to think of the coins as acquiring their nodes' Nimbers; moving a coin changes its Nimber. The effect on the Nim-sum of moving a coin is to add the Nimbers at the coin's starting and stopping nodes. Without going into the theory behind it, we simply note that a configuration of coins on the board is a losing position (for the first player) if and only if the coins' Nim-sum is 0. That is, if the Nim-sum is 0, then all moves produce a configuration with nonzero Nim-sum, while if the Nim-sum is nonzero, then at least one move leads to a configuration with Nim-sum 0.

This agrees with what we already observed for Puzzle One: No matter which node you put the four coins on, their Nim-sum is $a + a + a + a =$

$(a + a) + (a + a) = 0 + 0 = 0$, a losing position. But it also shows how to solve Puzzle Two. The Nim-sum of the nodes is

$$
\begin{aligned}
& 0 + 0 + 0 + 0 + 0 + 1 + 1 + 1 + 1 + 2 + 2 + 2 + 3 + 3 + 4 \\
= \; & (1 + 1) + (1 + 1) + (2 + 2) + 2 + (3 + 3) + 4 \\
= \; & 6,
\end{aligned}
$$

which is nonzero, hence the configuration is a first-player win. To change the Nim-sum to 0, it's necessary to move a coin in such a way that its starting and stopping nodes sum to 6. But the only available such sum is 4+2, corresponding to moving the coin from Jordan to Holcomb. This is the only winning move.

References

[1] Elwyn R. Berlekamp, John H. Conway, and Richard K. Guy, *Winning Ways*, second edition. A K Peters, Ltd., Natick, MA (2001).

[2] Thomas H. Cormen, Charles E. Leiserson, Ronald L. Rivest, and Clifford Stein, "Topological sort", Section 22.4 of *Introduction to Algorithms*, 2nd edition, MIT Press, Cambridge, MA (2001), pp. 549–552.

[3] Aviezri S. Fraenkel, Combinatorial Games, in: Proceedings of Symposia in Applied Mathematics (R.K. Guy ed.), American Mathematical Society, Providence RI, Volume 43(1991), 111-153.

[4] Richard K. Guy, *Fair Game: How to Play Impartial Combinatorial Games*. COMAP, Inc., Arlington, MA (1989).

[5] Richard K. Guy, Impartial Games, in: Proceedings of Symposia in Applied Mathematics (R.K. Guy ed.), American Mathematical Society, Providence RI, Volume 43(1991), 35-55.

Vive Recreational Mathematics

Underwood Dudley

In the course of denouncing some of what he saw as the foolishnesses of current mathematics education, Martin Gardner wrote

> Aside from its jargon, another objectionable feature of the year-book is that its contributors seem wholly unaware that the best way to keep students awake is to introduce recreational material that they perceive as fun. [5]

He was, as usual, right. The central problem of mathematics education is how to get students *engaged* with the material. To benefit from mathematics, students must work at it. Recreational mathematics problems give a good way to entice them into doing that.

Some enticement *is* necessary. The authoritarian days of yesteryear, when teachers and textbooks of mathematics felt no need to justify their subject, have gone forever. No longer could George Chrystal include, as he did in his 1886 *Algebra, An Elementary Text Book* [3, p. 154], three solid pages of problems, 77 in all, headed with "Express the following as rational

---

Underwood Dudley edited the *College Mathematics Journal*, wrote *Mathematical Cranks*, and is very proud of having an Erdős number of 1.

fractions at their lowest terms." He gave no reason why students should want to do such a thing to, for example, number 38 on the list,

$$\frac{(b+c)^2 + 2(b^2 - c^2) + (b-c)^2}{(b^4 - 2b^2c^2 + c^4)\left\{\frac{1}{(b-c)^2} + \frac{2}{b^2-c^2} + \frac{1}{(b+c)^2}\right\}}\ ;$$

they were just to *do* it, no questions asked.

The answer, given in the back of the book—Chrystal generously provided answers to even-numbered problems too—is, of all things, 1. That almost makes the problem recreational, though most of the fun would have been had by the person who constructed it. Nevertheless, it is possible that students who persevered to the right answer felt a small glow of pleasure when they found that the author agreed with them. This is akin to the satisfactions provided by recreational mathematics.

The introduction to Chrystal's *Algebra* gave no reason whatsoever for studying the subject. I think that it never occurred to Chrystal that any discussion was necessary. That is no longer the case. Now the fashion is for mathematics to be "relevant." The authors of algebra textbooks seem to think that since they have to try to get students to learn this stuff *somehow*, they will tell them how important it is to their lives and careers. No longer do they have the confidence to assume that the subject is worthy of study for its own sake.

The difficulty with the claim of importance is that it is false. A text intended for use in colleges, now in its third edition, whose name and author I will through charity suppress, has in its introduction the following:

> This text aims to show that mathematics is useful to virtually everyone. I hope that users will complete the course with greater confidence in their ability to solve practical problems.

But the problems in the book are hardly practical. Here is one:

> An investment club decided to buy $9000 worth of stock with each member paying an equal share. But two members left the club, and the remaining members had to pay $50 more apiece. How many members are in the club?

Were this a practical problem, one that really needed to be solved, the method of solution would be to ask a member of the club how many members it has. The member will know. If the member responds with a conundrum like the textbook problem, the member should be beaten about the head until he or she promises to behave in a more civilized manner.

I could continue to quote non-practical practical problems from this text at great length, but space forbids. Another algebra text, whose name and author I can furnish on request, has in its introduction that it

> shows students the relationship of chapter concepts and job skills—with applications developed through interviews and market research in the workplace that insure relevance.

Open the text and you will find "applications" such as:

> In preparation for the 2002 Winter Olympic Games in Salt Lake City, several people decide to pool their resources and share equally in the $12000 expense of renting a four-bedroom house in Salt Lake City for two weeks. The original number of people who agreed to share the house changed after two people dropped out of the deal because they thought the house was too small. Those left in the deal must now pay an additional $200 each of the rental. How many were sharing the house?

The alert reader, and even some unalert ones, may notice that this is the same problem as the one about the investment club. Though the numbers are different, it is isomorphic, and exactly as practical. But if we are to have "relevant" problems about solving quadratic equations, this is what we are forced to use.

It would be better, and more honest, to present *recreational* quadratic equations problems. Such things exist. In less than two minutes of searching I was able to find a problem in Dudeney's *The Canterbury Puzzles* [4, problem 66] that involves solving $x^4 - 20x^2 = -37$. I was disappointed to see that Dudeney wrote,

> It is an elegant little puzzle, and furnishes another example of the practical utility, on unexpected occasions, of a knowledge of the art of problem-solving.

It seems that authors *must* have relevance.

How is it that authors can get away with their claims of relevance? I suppose it must be that students are trained to be docile and to put up with whatever teachers do. They know that what goes on in classrooms is irrelevant.

Here is one last "application" from yet another textbook:

> Through experience and analysis, the manager of a storage facility has determined that the function $s(t) = -3t^2 + 12t + 10$ models the approximate amount of product left in the inventory after t days from the last supply.

The problem is to determine when to send in another order. Of course no manager of a warehouse ("storage facility" I suppose sounded classier to the author) ever has to solve a quadratic equation to determine when to reorder. Especially not this one, since if you make a table of the amount of product on hand

t	0	1	2	3	4
Amount on hand	10	19	22	19	10

you see that the inventory *breeds*. Inventory does not usually behave in this manner. Properly managed, no reorder would ever be needed.

If algebra were all that relevant to life and work, textbook writers would have no trouble at all in including relevant problems. There would be *thousands* of them. They could just pick any job and choose the first quadratic equation problem that they came to. That authors can't include them is evidence that there aren't any. Authors should stop trying to fool their students. They will be found out sooner or later, and some students may even see through the pretense right away and decide that mathematics is not worth learning. That is an error.

The purpose of having students solve algebra is to make them *use their heads*. The reason for teaching mathematics beyond arithmetic, and why it is now almost universal in the United States, is that it is by far the best way to teach people to reason. Mathematical reasoning leads from problem to solution, and the solution can be demonstrated to be correct. Mathematics illustrates better than any other subject the power of reason. Of course, we do not succeed in teaching everyone to reason, but there is no way to do the job any better. That is why the public supports mathematics education, even when they know that solving quadratic equations is not necessary on their jobs or in their lives.

Mathematics has always been learned by having students do artificial problems, those that both instructors and students understood have nothing to do with anything outside of mathematics. In the Rhind Papyrus, that collection of Egyptian problems compiled about 1650 BC, we find problem 40 [2]

> Give 100 loaves to five men so the shares are in arithmetic progression and the sum of the two smallest shares is 1/7 of the three greatest.

No one in ancient Egypt ever had to do that problem outside of a mathematics classroom (or wherever mathematics was taught in ancient Egypt). No one has ever needed to solve a problem like that in modern Egypt either, or anywhere else. The shares, by the way, are $1\frac{2}{3}$, $10\frac{5}{6}$, 20, $29\frac{1}{6}$, and $38\frac{1}{3}$.

The problem could be counted as recreational except that the Egyptians students to whom it was assigned probably did not get much enjoyment out of solving it. However, its unknown originator no doubt was very pleased to have discovered it, and derived great satisfaction when it was taken up by the mathematical establishment of the day and included in the curriculum.

But even the ancient Egyptian, not noted for their sense of fun, could not resist recreational mathematics. Problem 79 of the papyrus can be stated

> Seven houses, each with seven cats. Each cat with seven mice. Each mouse with seven ears of grain. Each ear of grain with seven hekats. How many?

(That last phrase means that when the seeds in an ear of grain were planted, they would yield seven hekats.) The answer is 19,607. Talk about adding apples and oranges! The ancient Egyptians were adding houses, mice, cats, and grain to get a perfectly meaningless answer. I have no doubt that the students of the papyrus found that problem much more appealing than those of the type "A quantity and its seventh amount to 19. What is the quantity?" with which the papyrus is plentifully supplied. Four thousand years ago, mathematics teachers knew the value of recreational mathematics. We should not ignore the wisdom of the past.

Problem 79 lives on, but with a twist, in the riddle about how many kits, cats, sacks, and wives were going to St. Ives. Recreational problems tend to live on because things that are valuable survive. That problem about how to get your wolf, your goat, and your cabbage across the river in a boat that would hold only two of them first appeared sometime around 800 AD. The monkey and coconuts problem, though not phrased in terms of monkeys and coconuts, dates back to around 850 AD. The snail in the well that climbs up three feet every day and slides back two feet every night has been at it since 1370. The longevity of such problems shows that they fill a need.

Of course not all recreational problems are ancient. The genre whose first example, by H. E. Dudeney, is SEND + MORE = MONEY can be dated exactly to July 1924, when it appeared in the *Strand* magazine. The field has undergone great development in the nineteenth and twentieth centuries. It is very healthy, which also shows that it fills a need.

The first recreational mathematics book was Bachet's *Problèmes plaisants et délectables qui se font par les nombres* [1] (in free translation, *Nice Number Problems*), published in 1612, the same year as the birth of logarithms. In it we find many familiar problems, including the one about having eight pints of wine that needs to be divided into two equal parts using three containers that hold 8, 5, and 3 pints. Everyone knows how to do that, but Bachet gives a harder example, dividing 16 pints into equal parts using containers of capacity 16, 11, and 6 pints. Bachet could do it in 14 steps. He also gives Tartaglia's generalization of the wolf, goat, and cabbage problem, the one about the three couples who needed to cross a river—perhaps the same one across which wolves had been ferried for hundreds of years—in a boat that could hold at most three people, where no woman could be in the presence of another man unless her husband is present. The women do most of the rowing.

Bachet's book starts slowly, as befits a pioneering work. Its first problem is a find-the-number game. The subject is to pick a number, triple it, and tell the game master whether the result is odd or even. If even, divide by two; if odd, add one and divide by two. Then triple again and announce how many times 9 goes into the result. The number first thought of is twice the result if the original number was even or twice the result plus one if it was odd.

It is not hard to see why this works. It is a natural for pedagogical application. Students could be challenged to find out why it works. They could then be given more complicated examples. They could make up their own. Group work and collaborative learning are currently all the rage, so up-to-date teachers could divide a class into groups, have each group devise a new way to find a number, and see if other groups could figure out why the method works. Reports could be written. Presentations could be made. Educational possibilities abound throughout recreational mathematics and, as Martin Gardner said, they should be exploited. It might be going too far to introduce actual courses called Introduction to Recreational Mathematics—a good way to suck the life out of a subject is to make a course out of it—but there are opportunities at all levels, in almost all classes.

We all know that recreational mathematics is fun, and that is why we participate in it. The reason why it is fun has not been deeply investigated, but the pleasure that people get from solving puzzles seems to be part of human nature. It may be that solving problems gives us the feeling that we are more in control of our environment, or that we have beaten back, a little, the borders of the unknown and thereby reduced the magnitude of the primordial chaos. It may reinforce our faith, misplaced or not, that we live in an orderly universe.

Besides the fun, for whatever reason it arises, recreational mathematics may have another use. In 1956, Theodore L. Shaw, an author today almost completely forgotten, published his *Precious Rubbish* [6], a criticism of art criticism. ("Rubbish" referred to the writings that art critics produced, and "precious" to their affected manner.) In it he set forth an idea that was new to me. It may be that it is a commonplace in the circles in which it is a commonplace, but I have not encountered it elsewhere. It is that the purpose of art is to *prolong life* in those who consume it. No scientific evidence for this assertion was presented, and it is hard to see how any could be produced, but it sounds reasonable. Lives that are nasty and brutish are likely to be short.

Can anyone deny that recreational mathematics is one of the arts? It is good to know that by participating in it we are increasing our longevity, quite independent of its ability, by stimulating us to use our minds, of staving off Alzheimer's disease.

Vive recreational mathematics!

References

[1] Claude-Gaspar Bachet, *Problèmes plaisants et délectables qui se font par les nombres*, 5th. ed. edited by A. Labosne, Blanchard, Paris, 1959.

[2] Arnold B. Chace, with the assistance of Henry P. Manning, *The Rhind Mathematical Papyrus*, 2 vols., Mathematical Association of America, Washington DC, 1927.

[3] G. Chrystal, *Algebra*, Adam and Charles Black, Edinburgh, 1886.

[4] Henry E. Dudeney, *The Canterbury Puzzles*, Dover edition, New York, 1958.

[5] Martin Gardner, Fuzzy new new math, *New York Review of Books* (September 24, 1998), reprinted in *Gardner's Workout*, A K Peters, Natick, MA, 2001, p. 303. The yearbook that Gardner found objectionable was Janet Trentacosta and Margaret J. Kenney, eds., *Multicultural and Gender Equity in the Mathematics Classroom*, National Council of Teachers of Mathematics, 1997.

[6] Theodore L. Shaw, *Precious Rubbish*, Stuart Art Gallery, Boston, 1956.

Gödelian Puzzles

Raymond Smullyan

When people ask me what Gödel's theorem is all about, I usually explain that about the turn of the century appeared two famous mathematical systems which appeared so comprehensive that every true mathematical statement could be proved in each of them. In 1931, to everyone's surprise, Kurt Gödel showed that this was not the case, that in each of the systems, and for a significant variety of related systems, there was a sentence which was neither provable nor refutable (disprovable) in the system—a sentence which was in fact true but not provable from the axioms of the system. Thus, the axioms were simply insufficient to determine which sentences were true and which were not.

The method by which Gödel achieved this was to assign to each sentence of the system a (positive whole) number, subsequently called the *Gödel number* of the sentence. He then constructed an extremely ingenious sentence G which asserted that a certain number n was the Gödel number of a sentence that was *not* provable in the system. The ingenious thing is that this number n was the Gödel number of the very sentence G! Thus, G asserted that its own Gödel number was the Gödel number of an *unprovable* sentence of the system, which is tantamount to G asserting its own non-provability in the system. This means that either G is true but

Raymond Smullyan is a master of recreational mathematics and logic puzzles, having authored many books on these subjects. Many years ago, he supported himself as a magician when both he and Martin Gardner were fellow students at the University of Chicago.

not provable in the system (as G asserts), or that G is false but provable
in the system (contrary to what G asserts). The latter alternative is ruled
out by the fact that it is obvious from the nature of the system that no
false sentences are provable in the system, so the first alternative must
hold—Gödel's sentence G is true but not provable in the system.

I have devised many recreational logic puzzles to illustrate Gödel's
method. Some of them take place on the island of *Knights and Knaves*,
where everything said by a knight is true and everything said by a knave is
false. Each inhabitant is either a knight or a knave. A logician once visited
the island. He was a hundred percent accurate in his beliefs and proofs:
he could never prove a false statement and he could prove only true ones.
The logician met a native of the island named Jal who made a statement
from which it follows that Jal must be a knight but it was impossible for
the logician to ever prove that he was a knight. What statement would
work? [**Puzzle 1**]

In my doctoral dissertation of 1959, one of the things I did was to
show how a sentence could be constructed that asserts not its own non-
provability, as Gödel's sentence did, but rather its own *refutability* (provability of its negation). Such a sentence must be false but not refutable
(and hence its negation must be true but not provable). The knight-knave
analogue is this: The same logician meets another native named Hal who
makes a statement from which it follows that he must be a knave, but
the logician can never prove that he is. What statement would work?
[**Puzzle 2**]

Here are some more Gödelian puzzles: On this same island, some of
the knights are titled *certified* knights and some of the knaves are titled
certified knaves. What statement could a native make that would imply
that he is an uncertified knight? [**Puzzle 3**] What statement could he make
that would imply that he is a certified knave? [**Puzzle 4**] What statement
could he make that would imply that he is either an uncertified knight or
a certified knave, but there is no way of telling which? [**Puzzle 5**] What
statement could a native make that would imply that he must be certified,
but there is no way of telling whether he is a knight or a knave? [**Puzzle 6**]

I sometimes introduced logic students to Gödel's theorem as follows: I
place a penny and a quarter on the table and explain to a student that he
is to make a statement and if the statement is true, then I promise to give
him one of the coins, not saying which one, but if the statement is false,
then I give him neither coin. What statement could the student make such
that to keep my word, I would have no choice but to give him the quarter?
[**Puzzle 7**]

Actually, in doing what I just described, I left myself wide open for
losing an arbitrary amount of money! If the student had thought of it,

he could have made a statement such that the only way I could keep to the agreement would be that I give him a million dollars. What statement would work? [**Puzzle 8**]

Craig's Gödelian Machine

How did Gödel ever manage to construct a sentence that asserted its own non-provability? Well, Inspector Craig of Scotland Yard, of whom I have written a great deal in many of my puzzle books, got intrigued with this. Although Craig is professionally a criminal investigator, he is equally interested in mathematics, logic, and philosophy.

When Inspector Craig heard of Gödel's construction, he decided to construct a logic machine that illustrates it with diabolical simplicity. His machine proved various sentences constructed from the following three symbols:

$$P \quad N \quad P^*$$

An *expression* is any string of these three symbols. For example, P*NPP is an expression, and so is NNPNP. Some of these expressions are called *sentences* and they have *meaning*, which will be explained shortly. In preparation for this, define the *repeat* of an expression X to be XX, i.e., X followed by itself. For example, the repeat of PNP is PNPPNP. Now, a sentence is any expression of one of the following four forms (where X is any expression whatsoever):

1. PX

2. NPX

3. P^*X

4. NP^*X

Now here is what the sentences mean:

1. The symbol P stands for *provable*, so that PX means that X is provable. Thus PX is *true* if and only if X is provable.

2. N stands for *not*, and thus NPX says that it is not the case that X is provable, or more briefly, that X is not provable. Thus NPX is *true* if and only if X is not provable.

3. P*X means that the *repeat* of X is provable. Thus P*X is *true* if and only if XX is provable. Note that P*X and PXX say the same thing.

4. NP*X means that the *repeat* of X is *not* provable. Thus NP*X is *true* if and only if XX is not provable.

We have here an interesting loop: Craig's machine is *self-referential* in that it proves various sentences that assert what the machine can and cannot prove. We are given that the machine is 100% accurate in that it proves only true sentences; it never proves any false ones. So, for example, if it proves PX, then X must really be provable. Conversely, if X is a provable sentence, it does *not* necessarily follow that PX must be provable. If X is provable, then PX must be true, but we were never given that all true sentences are provable, only that all provable sentences are true. In fact, there is a true sentence that is not provable by the machine, and your problem is to construct one. [**Puzzle 9**. Hint: Construct a sentence X that asserts that X is not provable.]

Interestingly enough, I discovered something that surprised even Inspector Craig. I discovered that it is possible in Craig's system to construct *two* sentences X and Y such that it must be the case that one of the two is true but not provable, but there is no way of telling which one it is! Can the reader find two such sentences? [**Puzzle 10**. Hint: Construct sentences X and Y such that X asserts that Y is provable and Y asserts that X is not provable. There are actually two different ways of doing this.]

I think it was puzzles of the last type that inspired the logician Melvin Fitting (formerly my student) to once introduce me at a mathematical meeting as: "I now introduce Professor Smullyan who will prove to you that either he doesn't exist or you don't exist, but you won't know which."

Solutions to Puzzles

1. A statement that works is "You cannot prove that I am a knight." If Jal were a knave, his statement would be false, which would mean that the logician *could* prove that Jal was a knight, contrary to the given condition that the logician never proves anything that is false. Therefore, Jal must be a knight; hence his statement must be true, which means that the logician can never prove that Jal is a knight.

2. A statement that works is "You can prove that I'm a knave." If the statement were true, then the native would have to be a knight (since knaves don't make true statements), and what he said would have to be the case, which would mean that the logician could prove

that the native is a knave, contrary to the given condition that the logician never proves false propositions. Thus, the statement must be false, and hence the native must be a knave. Furthermore, since the statement is false, it is *not* true that the logician can prove the native to be a knave. Thus, the native is a knave, but the logician can never prove that he is.

3. The statement "I am an uncertified knight." wouldn't work; a knave could say that. A statement that does work is "I am not a certified knight." A knave couldn't say that, because with a knave, it is *true* that he is not a certified knight. Thus, only a knight could say that, hence also it is true, so he must be an uncertified knight.

4. "I am an uncertified knave."

5. A statement that works is "I am uncertified." If he is a knight, then he is uncertified, as he claimed. If he is a knave, his claim is false, hence he is *not* uncertified, and so he is a certified knave.

6. A statement that works is "I am either a certified knight or an uncertified knave."

7. A statement that works is "You will not give me the penny." If the statement were false, that would mean that I *would* give him the penny, but the rule was that I wouldn't give him either coin for a false statement. Hence, the statement must be true, which means that I won't give him the penny. Yet, I must give him one of the coins for a true statement; hence I must give him the quarter.

   This problem occurred to me in relation to Gödel's construction as follows: I thought of the quarter as the analogue of *truth* and the penny as the analogue of *provability*, and so the sentence "You will not give me the penny" is the analogue of the Gödel sentence, which effectively says "I am not provable."

8. A statement that works is "You will give me neither the penny nor the quarter nor a million dollars." If the statement were true, then I would have to give him either the penny or the quarter as agreed, but my doing so would obviously falsify the statement that I will give him *none* of those three things. Thus, the statement must be false. Since it is false that I give him *none* of those three things, I must give him at least *one* of those three things, but I can't give him either the penny or the quarter for a false statement, so I have no choice other than to give him a million dollars.

9. A sentence that must be true but not provable in the system is NP*NP*. It asserts that it is not the case that the repeat of NP* is provable, but the repeat of NP* is the very sentence NP*NP*. Thus, NP*NP* is true if an only if it is not provable, which means that either it is true and not provable, or it is false but provable. The latter alternative is ruled out by the given condition that no false sentences are provable. Hence, NP*NP* is true but not provable in the system.

10. We want sentences X and Y such that X is true if and only if Y is provable, and Y is true if and only if X is not provable.

First suppose we had such sentences. Either X is true or X is false. Suppose X is true. Then Y really is provable, as X asserts. Since Y is then provable, it must be true (as only true sentences are provable), which means that X is not provable, as Y asserts. Thus, if X is true, it is not provable. Now, suppose that X is false. Then X *falsely* asserts that Y is provable; hence in reality Y is *not* provable. But also, since X is false, it is not provable; hence Y, which asserts this fact, must be true! Thus, if X is false, then Y is true but not provable.

In summary, if X is true, then X is true but not provable, and if X is false, then Y is true but not provable. There is no way to tell whether X is true or false; hence there is no way to tell whether it is X or Y that is the one which is true but not provable in the system.

Now we exhibit such sentences X and Y. One solution is to take $X = \text{PNP*PNP*}$ and $Y = \text{NP*PNP*}$. X is the sentence PY; hence X asserts that Y is provable. Y asserts that it is not the case that the repeat of PNP* is provable, but the repeat of PNP* is X itself.

Another solution is to take $X = \text{P*NPP*}$ and $Y = \text{NPP*NPP*}$. Clearly, Y says that X is not provable and X says that the repeat of NPP*, which is Y, is provable.

Part III

Brainteasers

A Bouquet of Brainteasers

Chris Maslanka

Puzzles

1. Roses. A bouquet contains red roses, white roses, and blue roses. According to the florist, the number of red roses and white roses comes to 100; the number of white roses and blue roses comes to 53. The number of blue roses and red roses come to less than that.

How many roses of each color are there?

2. Dropouts. Identify these two words in which each asterisk marks a missing letter:

$$B_*R_*A_*I_*N$$

$$_*B_*R_*A_*I_*N$$

3. Small Change. These two words differ only in their third letters. One word means fun and the other means serious thinking!

Chris Maslanka is a puzzlist, pianist, and presenter of radio programmes for the BBC.

$$_{**}\mathrm{L}_{*******}$$

$$_{**}\mathrm{R}_{*******}$$

What are the two words?

4. **Plus Fours.** "The time is four o'clock," announced Professor Turki as he put the petit-fours into the oven.

"We need at most seven fours to write the number 100 as a sum of numbers made up entirely of the digit 4:

$$44 + 44 + 4 + 4 + 4 = 100.$$

We need at most sixteen fours to make 1,000 in this way:

$$444 + 444 + 44 + 44 + 4 + 4 + 4 + 4 + 4 + 4 = 1,000."$$

What is the smallest number of fours we need to represent 1,000,000 in this way? What about 10^n?

5. **Couscous.** COUSCOUS is 40,000 short of being a perfect square. As such it is unique. What is COUSCOUS?

(Assume each of the letters of COUSCOUS stands for a digit in base 10. Different letters stand for different digits.)

6. **Six Easy Pieces.** Divide this figure into six pieces all the same size and shape:

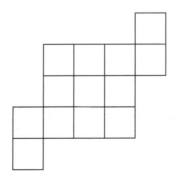

7. **Seventh Heaven.** Agnes loves olives. The tin she has found in the cupboard certainly contained green olives and black olives to begin with, but someone evidently got to it before Agnes did and ate a number of them. The first seven olives Agnes now removes at random from the tin all turn out to be green. The probability of this happening is exactly 50-50. Given that, what can you expect of the colors of the remaining seven olives? Hurry—they're going fast....

8. **Pardon My French.** Fill in the missing letters in this two-word phrase:

$$_{**}\text{LM NO}_{**}$$

9. **Cube-Ah!** If every packet of *Cube-Ah!* Castro Sugar retails for exactly 90 cents and each packet carries a voucher, nine of which may be exchanged for yet another packet of *Cube-Ah!*, what is the value of the contents of a packet? (Ignore the cost of the packaging.)

10. **X Words.** What uncommon property do all these words have in common?

ABLE DON OR ANT ON PIN

11. **Reciprocals.** The reciprocal 1/11 can be written as the sum of the reciprocals of two positive even numbers in two different ways. One way is of course 1/22 + 1/22. What is the other?

12. **Last but not least.** What letter is missing from this sequence:

T R S G M S ?

Solutions

1. Let the number of red roses be r, white w, and blue b. Then $r + w = 100$; $w + b = 53$; and $b + r = x$, where x is less than 53. Adding the three equations counts each sort of flower twice, so x must be odd. We know that x is less than 53. Because we are told that there is a blue flower, $x + 53$ must produce a total more than 100, since it counts not only the white flowers and the red flowers but also the blue flowers twice over. So x must be larger than 47. It could be 49 or 51. Adding the three equations and dividing by two shows that $r + w + b = 101$ in the first case. This leads to one blue

flower, 48 red flowers, and 52 white flowers. But we are told there are blue flowers in the plural. So x must equal 51. Then $r + b + w = 102$, yielding 2 blue flowers, 49 red flowers, and 51 white flowers.

2. BARBARIAN & ABERRATION.

3. CELEBRATION and CEREBRATION.

4. For 1,000,000, the number of fours required is 43. For 10^n, $n \geq 2$, the number of fours required is $9n - 11$. To see this, let N be the number of fours required to represent 10^n as a sum of strings of fours. Dividing everything by 4, N is also the number of ones required to represent $25 \times 10^{n-2}$ as the sum of strings of the digit 1 (repunits). Let's write 1_k for the decimal number represented by a string of k ones, i.e., $1 + 10 + 100 + \ldots + 10^{k-1}$. One way to write a number X as a sum of numbers of the form 1_k is to repeatedly add the largest 1_k that does not bring the sum beyond the desired total X. For $X = 25 \times 10^{n-2}$, this algorithm starts with $2 \times 1_n$ to get the first digit 2 and simultaneously two-fifths of the second digit 5. Then it adds $2 \times 1_{n-1}$ to get two more fifths of the second digit 5; adding a third 1_{n-1} would overshoot the desired sum. At this point all digits beyond the first two in the sum are fours. Then it adds $5 \times 1_{n-2}$ to raise these digits to 9, and finally it adds 1 to trigger all carries beyond the first two digits and raise the second digit from 4 to 5. In summary,

$$25 \times 10^{n-2} = 2 \times 1_n + 2 \times 1_{n-1} + 5 \times 1_{n-2} + 1$$

The number of ones used by this representation is $2n + 2(n - 1) + 5(n-2)+1 = 9n-11$. We claim that no representation uses fewer ones. First, the $n-2$ zeros in $25 \times 10^{n-2}$ must be formed by overflowing into carry, because the first two digits 25 are either formed by 1_n's and/or 1_{n-1}'s or by carrying from lower digits, and in either case all of the zero digits are "corrupted" with nonzero digits. Therefore the number of ones needed to form these zero digits is at least $9(n - 2) + 1 = 9n - 17$, nine per digit, to form a near-overflow condition, and one more to trigger a cascading carry. The zero digits would need more ones if part of the 9 in a digit is formed by a lower carry, or if a digit overflowed more than once. The first two digits 25 get one unit for free from the overflow of lower digits. The number of ones required to represent the remaining 24 is clearly $2 + 4 = 6$. Therefore the number of ones required is at least $9n - 11$, which is matched by our construction.

5. COUS = 9601, and we claim that this solution is unique. COUS-COUS = COUS × 10,001 = $X^2 - 40,000 = (X - 200)(X + 200)$. Note that 10,001 has prime factorization 73 × 137. These prime factors of 10,001 either divide the same factor of $X^2 - 40,000$ or divide different factors. Without loss of generality (by letting X be either positive or negative), either 10,001 divides $X - 200$ or else 73 divides $X - 200$ and 137 divides $X + 200$. In other words, either $X \equiv 200 \pmod{10{,}001}$ or else $X \equiv 200 \pmod{73}$ and $X \equiv -200 \pmod{137}$. In the former case, the possible values for X are ..., -19,802, -9,801, 200, 10,201, ..., but of these only -9,801 gives a meaningful value for COUS: $(-9{,}801)^2 - 40{,}000 = 9{,}601{,}9601$, i.e., COUS = 9601. $X = 200$ gives $X^2 - 40,000 = 0$, i.e., COUS = 0000, and all other choices for X give $X^2 - 40,000$ with more than eight digits. In the latter case, the Chinese Remainder Theorem (or the Euclidean Algorithm) implies $X \equiv 8{,}157 \pmod{10{,}001}$, so the possible values for X are ..., -11,845, -1,844, 8,157, 18,158, ... The only choices for which $(X^2 - 40,000) / 10,001$ gives a number with at most four digits is $X = 8{,}157$, which gives COUS = 6649, and $X = -1{,}844$, which gives COUS = 0336, but neither of these solutions has different digits for different letters as required.

6. The puzzle is certainly solvable if you regard the diagram as representing a net. Then it can be cut into six pieces each in the form of the digit 6. Some of these will look like 9s in the position they originally occupy in the net.

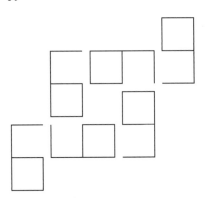

7. One is black and six are green. One "lazy" way of "seeing" this is to entertain the idea that if all the olives are green bar one, the chances of her removing the black one among the first seven must be the same as the changes of her removing it among the second seven, i.e., 50-50.

8. FILM NOIR.

9. 80 cents. Without the vouchers it would be 90 cents a packet. It may seem at first that for every nine packets you have enough vouchers for an extra packet, which suggests 81 cents per packet. But this packet has a voucher on it as well, which is worth something, so 81 cents cannot be right.

Run the following thought experiment: Borrow one packet from a friend. Now buy eight packets for $90 \times 8 = 720$ cents. Consume nine packets, and give back to the lender the free packet that you get with the vouchers. You have had nine packets at the cost of 720, which is 80 cents a packet.

10. All of the words can be preceded by TEN to make other words.

11. $1/12 + 1/132$. In general, for a prime p, $1/p = 1/(p+1) + 1/p(p+1)$.

12. E. The letters in the sequence are the last letters of the words in the question.

Sliding-Coin Puzzles

Erik D. Demaine and Martin L. Demaine

In what ways can an arrangement of coins be reconfigured by a sequence of moves where each move slides one coin and places it next to at least two other coins? Martin Gardner publicized this family of sliding-coin puzzles (among others) in 1966. Recently, a general form of such puzzles was solved both mathematically and algorithmically. We describe the known results on this problem, and show several examples in honor of Martin Gardner for the Fifth Gathering for Gardner.

Puzzles

Sliding-coin puzzles ask you to rearrange a collection of coins from one configuration to another using the fewest possible moves. Coins are identical in size, but may be distinguished by labels; in some of our examples, the coins are labeled with the letters G, A, R, D, N, E, R. For our purposes, a *move* involves sliding any coin to a new position that touches at least two other coins, without disturbing any other coins during the motion.

The rest of this section presents several sliding-coin puzzles.

The Demaines are a father-and-son team. **Erik D. Demaine**, the father, is an assistant professor of computer science at MIT who likes anything algorithmic or geometric. **Martin L. Demaine**, the son, is a researcher in computer science at MIT who likes anything mathematical and artistic.

Triangular Lattice

We start with some basic puzzles that are *on the triangular lattice* in the sense that the center of every coin is at a vertex of the planar lattice of equilateral triangles. The restriction that a move must bring a coin to a new position that touches at least two other coins forces a puzzle to stay on the triangular lattice if it is originally on it.

TRIANGULAR LATTICE PUZZLES

Classic Puzzles

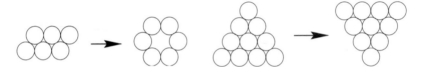

Puzzle 1. Rhombus to a circle (3 moves).

Puzzle 2. Turn the pyramid upside-down (3 moves).

New Puzzles

Puzzle 3. Pyramid to a line (7 moves). Source unknown.

Puzzle 4. Spread out the GARDEN (9 moves).

Square Lattice

Next we give a few puzzles *on the square lattice*. Here the centers of the coins are at vertices of the planar lattice of squares, and we make the additional constraint that every move brings a coin to such a position. The restriction that a move must bring a coin to a new position that touches at least two other coins does not force the puzzle to stay on the square lattice, but this additional constraint does.

SQUARE LATTICE PUZZLES

Penta Puzzles

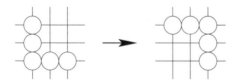

Puzzle 5. Flip the L (4 moves).

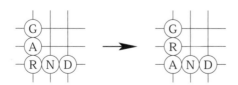

Puzzle 6. Rotate the L by 90° (8 moves).

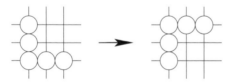

Puzzle 7. Fix the spelling of GRAND (8 moves).

Hard Puzzles

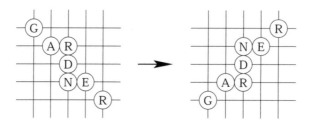

Puzzle 8. Flip the diagonal (hard).

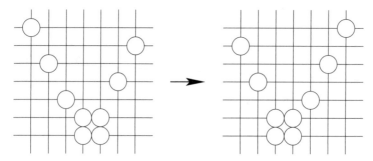

Puzzle 9. Flip the roman numeral V (very hard).

History

Sliding-coin puzzles have long been popular. For example, the classic Puzzle 1 is described in Martin Gardner's *Scientific American* article on "Penny Puzzles" [8], in *Winning Ways* [1], in *Tokyo Puzzles* [7], in *Moscow Puzzles* [9], and in *The Penguin Book of Curious and Interesting Puzzles* [12]. Langman [10] shows all 24 ways to solve this puzzle in three moves. Puzzle 2 is another classic [2, 7, 8, 12]. Other puzzles are presented by Dudeney [6], Fujimura [7], and Brooke [4].

The historical puzzles described so far are all on the triangular lattice. Puzzles on the square lattice appear less often in the literature but have significantly more structure and can be more difficult. The only published example we are aware of is given by Langman [11], which is also described by Brooke [4], Bolt [3], and Wells [12]; see Puzzle 10. However, the second of these puzzles does not remain on the square lattice; it only starts on the square lattice, and the only restriction on moves is that the new position of a coin is adjacent to at least two other coins.

Puzzle 10. Rearrange the H into the O in four moves while staying on the square lattice (and always moving adjacent to two other coins), and return to the H in six moves using both the triangular and square lattices.

Mathematics

A paper by Helena Verrill and the present authors [5] solves a large portion of the general sliding-coin puzzle-solving problem: given two configuration of coins, is it possible to re-arrange the first configuration into the second via a sequence of moves? One catch is that, for the results to apply, a move must be redefined to allow a coin to be picked up and placed instead of just slid on the table. Another catch is that the solution does not say anything about the minimum number of moves required to solve a specific puzzle, though it does provide a polynomial upper bound on the number of moves required. (From this information we can also determine which puzzle requires roughly the most moves, among all puzzles.) Despite these catches, the results often apply directly to sliding-coin puzzles and tell us whether a given puzzle is solvable, and if so, how to solve it. The ability to tell whether a puzzle is solvable is ideal for puzzle design.

A surprising aspect of this work is that there is an efficient algorithm to solve most sliding-coin puzzles, which runs fast even for very large puzzles. In contrast, most other games and puzzles, when scaled up sufficiently large, are computationally intractable.

Triangular Lattice

It turns out to be fairly easy to characterize which triangular-lattice puzzles are solvable. Part of what makes this characterization easy is that most puzzles are solvable. Consider a puzzle with an initial configuration that differs from the goal configuration. There are a few basic restrictions for this puzzle to be solvable:

1. There must be at least one valid move from the initial configuration.

2. The number of coins must be the same for the initial and goal configurations.

3. At least one of the following four conditions must hold:

  (a) The final configuration contains a triangle of three mutually touching coins.

  (b) The final configuration contains four connected coins.

  (c) The final configuration contains three connected coins and two different touching coins (as in Puzzle 4).

  (d) The puzzle is solvable by a single move.

4. If the coins are labeled and there are only three coins (a rather boring situation), then the goal configuration must follow the same three-coloring of the triangular lattice.

After some thought, you will probably see why each of these conditions must hold for a puzzle to be solvable. What is more surprising, but beyond the scope of this article, is that these conditions are enough to guarantee that the puzzle is solvable. Interested readers are referred to [5] for the proof.

Square Lattice

Solvable square-lattice puzzles are trickier to characterize. Much more stringent conditions must hold. For example, it is impossible for a configuration of coins to get outside its enclosing box. This property is quite different from the triangular lattice, where coins can travel arbitrary distances.

A notion that turns out to be particularly important with square lattices is the *span* of a configuration. Figure 1 shows an example. Suppose we had a bag full of extra coins, and we could place them onto the lattice at any empty position adjacent to at least two other coins. If we repeat this process for as long as possible, we obtain the span of the configuration.

 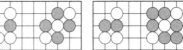 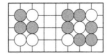

Figure 1. In general, the span consists of one or more rectangles separated by distance at least two empty spaces.

A key property of span is that it can never get larger by a sequence of moves. The span effectively represents all the possibly reachable positions in a configuration. So if you are to move the coins from one configuration to another, the span of the first configuration better contain the span of the second configuration.

This condition is not quite enough, though. In fact, the exact conditions are not known for when a square-lattice puzzle can be solved. However, most solvable configurations have some *extra coins* whose removal would not change the span of the configuration. The main result of [5] says that if a configuration has at least two extra coins, then it can reach any other configuration with the same or smaller span.

This result is complicated, leading to some puzzles with intricate solutions, as in Puzzles 8 and 9, for example. To work your way up to these difficult puzzles, we have provided some "warm up" puzzles that involve five coins, in the spirit of the "penta" theme of the Fifth Gathering for Gardner.

Solutions

Solutions to Puzzles 1, 2, 3, 5, 6, and 7 were found by an exhaustive breadth-first search, and as a result we are sure that the solutions use the fewest possible moves. For Puzzle 4, it is conceivable that using more than two "rows" leads to a shorter solution; the solution below is the best among all two-row solutions. Puzzles 8 and 9 are left as challenges to the reader; we do not know the minimum number of moves required to solve them. The second half of Puzzle 10 was solved by hand, but the number of moves is minimum: the maximum overlap between the H and the O is four coins, three moves are necessary to enter the triangular lattice and return to the square lattice, and every move starting with Move 3 puts a coin in its final position.

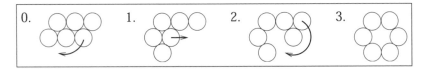

Solution to Puzzle 1. Three moves.

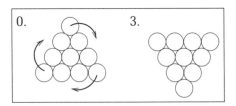

Solution to Puzzle 2. Three moves.

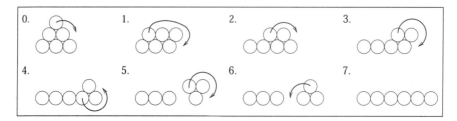

Solution to Puzzle 3. Seven moves.

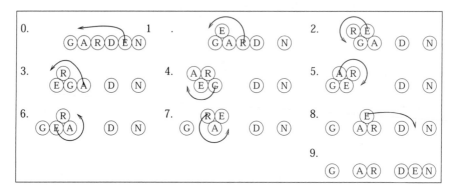

Solution to Puzzle 4. Nine moves.

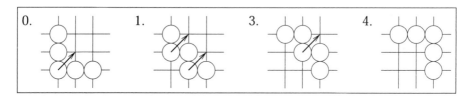

Solution to Puzzle 5. Four moves.

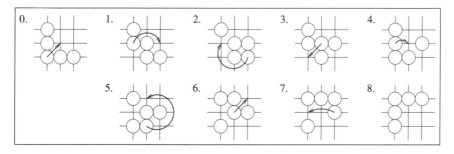

Solution to Puzzle 6. Eight moves.

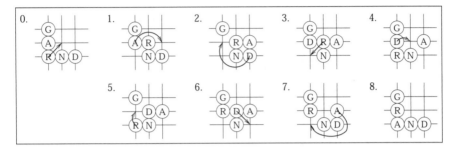

Solution to Puzzle 7. Eight moves.

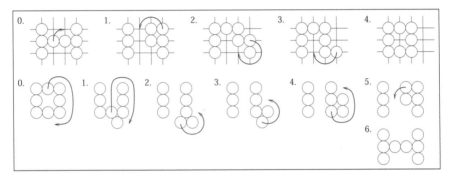

Solution to Puzzle 10. Four moves there and six moves back.

References

[1] Elwyn R. Berlekamp, John H. Conway, and Richard K. Guy. A solitaire-like puzzle and some coin-sliding problems. In *Winning Ways for Your Mathematical Plays*, Volume 2, pages 755–756. Academic Press, London, 1982.

[2] Brian Bolt. Invert the triangle. In *The Amazing Mathematical Amusement Arcade*, amusement 53, page 30. Cambridge University Press, Cambridge, 1984.

[3] Brian Bolt. A two touching transformation. In *Mathematical Cavalcade*, puzzle 20, page 10. Cambridge University Press, Cambridge, 1991.

[4] Maxey Brooke. *Fun for the Money.* Charles Scriber's Sons, New York, 1963. Reprinted as *Coin Games and Puzzles* by Dover Publications, New York, 1973.

[5] Erik D. Demaine, Martin L. Demaine, and Helena Verrill. Coin-moving puzzles. In R. J. Nowakowski, editor, *More Games of No Chance*, pages 405–431. Cambridge University Press, 2002. Collection of papers from the MSRI Combinatorial Game Theory Research Workshop, Berkeley, California, July 24–28, 2000. http://www.arXiv.org/abs/cs.DM/0204002.

[6] Henry Ernest Dudeney. "The four pennies" and "The six pennies". In *536 Puzzles & Curious Problems*, problems 382–383, page 138. Charles Scribner's Sons, New York, 1967.

[7] Kobon Fujimura. "Coin pyramids," "Four pennies," "Six pennies," and "Five coins". In *The Tokyo Puzzles*, puzzles 23 and 25–27, pages 29–33. Charles Scribner's Sons, New York, 1978.

[8] Martin Gardner. Penny puzzles. In *Mathematical Carnival*, chapter 2, pages 12–26. Alfred A. Knopf, New York, 1975. Appeared in *Scientific American*, 214(2):112–118, February 1966, with solutions in 214(3):116–117, March 1966.

[9] Boris A. Kordemsky. A ring of disks. In *The Moscow Puzzles*, problem 117, page 47. Charles Scribner's Sons, New York, 1972.

[10] Harry Langman. Curiosa 261: A disc puzzle. *Scripta Mathematica*, 17(1–2):144, March–June 1951.

[11] Harry Langman. Curiosa 342: Easy but not obvious. *Scripta Mathematica*, 19(4):242, December 1953.

[12] David Wells. "Six pennies," "OH-HO," and "Inverted triangle". In *The Penguin Book of Curious and Interesting Puzzles*, puzzle 305, 375, and 376, pages 101–102 and 125. Penguin Books, New York, 1992.

Underspecified Puzzles

David Wolfe and Susan Hirshberg

An underspecified puzzle is one that appears to have multiple solutions. The puzzles here do not even have any instructions, so that part of each puzzle is to figure out what the puzzle is. What is surprising is that, when you've found the intended solution, we suspect you'll know. Some underspecified puzzles are culturally specific, making them difficult or impossible to solve if you're not familiar with the relevant culture; our apologies. There is one puzzle on each line below.

↑

____ ____ ____ ____ ____ ____ ____ ____ ____ ____ ____ ____

____ ____ ____ ____ ____ 30030 . . . .

____ ____ ____ ____ 8/5 . . .

4 ____ ____ 7 ____ ____ . . . .

____ ____ ____ ____ 50 ____

____ ____ ____ ____ 50 ____

David Wolfe (along with Tom Rodgers) edited the last Gardner volume *Puzzlers' Tribute: A Feast for the Mind*, and co-authored *Mathematical Go: Chilling Gets the Last Point* with Elwyn Berlekamp. **Susan Hirshberg** remains conspicuously obscure.

P   P   P   ____   P   ____   P   P

____ ____ ____ ERE   ____ ____ ____

H   ____ ____ ____ ____ ____ ____ .... ____

!   ____ ____ ____ ____ ____ ____ ____ ____ ____

R   ____ ____ ____ ____ ____ ____   R

LOVES HITS PROSE HEWN SHEET SETTLER EAR STORED GIRTH

Solutions

- Clock faces, 1:00 through midnight.

- 2, $2\cdot3$, $2\cdot3\cdot5$, $2\cdot3\cdot5\cdot7$, $2\cdot3\cdot5\cdot7\cdot11$, $2\cdot3\cdot5\cdot7\cdot11\cdot13$, ...

- $1/1$, $2/1$, $3/2$, $5/3$, $8/5$, $13/8$, ... (successive approximations to the golden ratio as a ratio of consecutive Fibonacci numbers)

- 4 score and 7 years ago ...

- U.S. coins: 1, 5, 10, 25, 50, 100

- U.S. bills: 1, 5, 10, 20, 50, 100

- P P P A P O P P (Peter Piper picked a peck of pickled peppers)

- Able was I ere I saw Elba

- Periodic table

- Characters over the numbers on a typewriter or keyboard

- Starting position of chess pieces on the first rank of a chessboard

- Solve this poser when these letters are sorted right

Acknowledgments

The first puzzle was composed by Martin Demaine.

A Cryptic Crossword Puzzle in Honor of Martin Gardner

Robert A. Hearn

About Cryptic Crosswords

Most Americans have not been exposed to cryptic crosswords, although they are the more popular form of crossword in England. Perhaps the most visible difference to those familiar with convential crosswords is that cryptic crosswords have substantially more black squares. The more substantial difference is in the clues. In a conventional crossword, a clue is a simple definition or synonym of the answer. In a cryptic crossword, the clue contains a definition, but it also contains some wordplay giving another route to the answer. Thus, each clue provides an independent, hopefully entertaining, word puzzle.

The main challenge in cryptic crosswords lies in figuring out how to parse the clues. Here is a simple example. The (6) following the clue means that the answer has six letters; a (5, 3) would indicate a five letter word followed by a three letter word.

> Don't succumb to Twisted Sister. (6)

The answer is "resist." If you've never done a cryptic crossword before, this will not be obvious. The tricky thing about the clues is that the

---

Bob Hearn cowrote the Macintosh program ClarisWorks (now called AppleWorks). He is currently studying artificial intelligence at MIT.

wording and punctuation often lead you away from the answer, and towards an unrelated, whimsical interpretation. Once the above clue is more elaborately punctuated, the answer becomes less mysterious.

Let's adopt the following "extended punctuation" conventions: parentheses are used for grouping, and square brackets are used to enclose words that constitute the definition or synonym. This will become clearer below. Here is the newly punctuated version of our clue:

[Don't succumb to] (twisted "sister") (6)

This punctuation makes it apparent that "don't succumb to" is the definition, and the wordplay consists of twisting (i.e., forming an anagram of) the letters of "sister."

Usually the definition occurs at either the beginning or the end of the clue, so the first thing to do when you see a clue is to try to pick out the definition. Then look for cue words indicating the kind of wordplay involved. Anagrams are common—look for words like "broken," "scrambled," "disturbed," "out," etc. But there are many other kinds of wordplay allowed, and often a single clue will contain multiple kinds of wordplay.

Here is a more complicated example, which doesn't use an anagram, appropriately punctuated:

([Shady scheme:] "way") to become [a mathematician] (6)

The answer is "Conway," i.e., a mathematician; the substitution for shady scheme is "con." Note that, as with conventional crosswords, substitutions can yield more specific words: [state] can be "Texas." However, [Texas] can't be "state," but [Texas, e.g.] could be.

Another form of wordplay involves picking out pieces of words or phrases. E.g., [the heart of Texas] could be "x" (also "exa"); [some fish] indicates merely that some of the letters of "fish" should be used; [the head of state] is "s." Sometimes the entire answer is literally spelled out, hidden in the clue:

[Edge] (in "the middle") (3) = HEM

Sometimes the wordplay will consist merely of a second definition, using a different sense of the word:

[Canine] [speaker] (6) = WOOFER

In addition to being scrambled or picked out, letters can be reversed: backward could be read as back "ward" = "draw." In general, any word that makes sense as an operation to apply to a string of letters can be

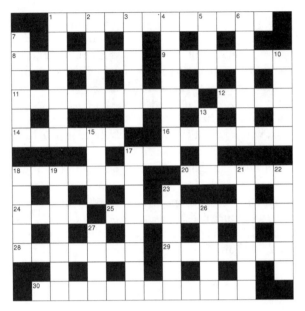

Across

1 In the cinema, thematic allegory is rarely precise. (12)
8 Abruptly start east, side by side. (7)
9 Tangle, tangle, tangle—finish with grace. (7)
11 Lunatic incites officer to retract offer—not quite rational. (10)
12 Idly alter end of poem. (4)
14 Swallow most of cocktail. (6)
16 A tall, hollow insect: get one in Georgia. (7)
17 Enjoyment ends if you frown. (3)
18 Fish dinner, without marlin tail? I'm a skeptic. (7)
20 Sly mob defaced sign. (6)
24 Plans for bouncing spam. (4)
25 Saint from Malta to lead traveler who's lost faith. (5, 5)
28 Actions intended to elicit a response strangely limit us. (7)
29 Time to ascend crest—take out notes, divide into three. (7)
30 Helga, an ex-fox, needs work to become a flexible model. (12)

Down

1 Top off Perrier with a bit of rum—that's beginning to get more festive. (7)
2 Educate retinue. (5)
3 Imply ailment ends, then begins. (6)
4 Nationality of a man harboring Greenland's founder. (8)
5 Thought that's four-fifths perfect. (4)
6 A boy with some dingy lamp! (7)
7 Occurs throughout first presentation; assimilate half. (6)
10 Crushed talus in oil-producing state. (5)
13 Pay about 50, as you like it, perhaps. (4)
15 Some prions found in plasma. (4)
17 Ability to choose document that's priceless? (4, 4)
18 For amusement, go risk life. (5)
19 Alligator could be violent—repel it. (7)
21 Catcall comes before cant: true or false? (7)
22 Lion is mighty surly, at first—it's in restraints. (6)
23 Search all over Texas for mountain summit. (6)
26 One caught in heartless, flabbergasting affair. (5)
27 Left an Italian in orbit. (4)

used that way. When combined with the rule that substitutions can be made arbitrarily, this is theoretically all you need to know to solve cryptic crosswords. However, there are a lot of standard patterns and substitutions. Some of the more frequent substitutions are Roman numerals ([six] =

"VI"), compass directions ([point] = "N," "S," "E," ...), foreign articles
(["the" French] = "le"), and common abbreviations.

There are two special punctuation conventions that deserve mention.
When a clue ends with a question mark, it often flags a pun, or an outrageous substitution. When a clue ends with an exclamation mark, it signals
what's called an "and literal" clue: instead of being entirely separate, the
normal definition and the wordplay then overlap completely—the clue can
be read both as a normal definition, and literally as instructions for forming
the word.

Finally, be warned that many clues use several of the above techniques,
all at once!

For more examples of cryptic crossword clues, see http://www
.theatlantic.com/unbound/wordgame/cluesolv.htm. The "extended punctuation" conventions used here are not standard, but seem to me to be a
reasonable way to explain clues (and to justify answers!).

WARNING—the answers follow.

Answers

Across

1 (In "the cinema thematic allegory is rarely") [precise] =
   MATHEMATICAL

8 (("Abruptly" start) "east") [side by side] = ABREAST

9 (Tangle ("tangle" ("tangle" finish))) [with grace] = ELEGANT

11 ((Lunatic "incites") ("officer" to retract ("offer" not quite)))
   [rational] = SCIENTIFIC

12 ("Idly" alter end of) [poem] = IDYL

14 [Swallow] (most of [cocktail]) = MARTIN

16 ("A" ("tall" hollow) [insect] get [one]) [in Georgia] = ATLANTA

17 [Enjoyment] (ends "if" "you" "frown") = FUN

18 ([Fish] ("dinner" without ("marlin" tail))) [I'm a skeptic] =
   GARDNER

20 ("Sly mob" defaced) [sign] = SYMBOL

24 [Plans] for (bouncing "spam") = MAPS

25 ([Saint] "from" ("Malta" to lead)) [traveler who's lost faith] =
PETER FROMM

28 [Actions intended to elicit a response] (strangely "limit us") =
STIMULI

29 ([Time] [to ascend] ("crest" take out [notes])), [divide into
three] = TRISECT

30 ("Helga an ex-fox" needs work) to become [a flexible model] =
HEXAFLEXAGON

Down

1 ((Top off "Perrier") with (a bit of "rum") that's beginning) to get
[more festive] = MERRIER

2 [Educate] [retinue] = TRAIN

3 [Imply] ("ailment" ends then begins) = ENTAIL

4 [Nationality] of ("a man" harboring [Greenland's founder]) =
AMERICAN

5 [Thought] that's (four-fifths [perfect]) = IDEA

6 [A boy with some dingy lamp] , ("A" [boy] with some "dingy
lamp") = ALADDIN

7 [Occurs throughout] ((first "presentation") ("assimilate" half)) =
PASSIM

10 (Crushed "talus") [in oil-producing state] = TULSA

13 ("Pay" about [50]) [as you like it, perhaps] = PLAY

15 (Some "prions") [found in plasma] = IONS

17 [Ability to choose] ([document] that's [priceless?]) =
FREE WILL

18 [For amusement] [go risk life] = GAMES

19 [Alligator, could be] (violent "repel it") = REPTILE

21 ([Catcall] comes before [cant]) [true or false] = BOOLEAN

22 (("Lion" "is" "mighty" "surly" at first) "it" 's in) [restraints] = LIMITS

23 (Search "all over Texas") for [mountain summit] = VERTEX

26 ([One] caught in (heartless "flabbergasting")) [affair] = FLING

27 ([Left] [an Italian]) [in orbit] = LUNA

Absolute Martin

J. Carey Lauder

Martin Gardner's book *Encyclopedia of Impromptu Magic* lists tricks from A to Z. However, there are no tricks for five letters of the alphabet. In my photo (see Color Plate VI), there are 21 different items, some single, some multiple, each beginning with a letter of the alphabet for which Martin's book includes a trick. There are apples for A, bottle tops for B, etc.—you get the idea. Which five letters were left off the list? (The answer is given on the next page.)

ABSOLUTE MARTIN

Carey Lauder is a full-time commercial photographer in Winnipeg, Canada and has been involved in magic for the past 12 years. He became involved with the Gathering due to his friendships with Mel Stover and Harry Eng.

Answer

The omitted letters are J, Q, U, X, and Y. The represented letters are as follows:

A	=	apples
B	=	bottle tops
C	=	corks
D	=	dice
E	=	eggs
F	=	fork
G	=	glove
H	=	handkerchief
I	=	ice cubes
K	=	key
L	=	life savers
M	=	matches
N	=	nuts
O	=	olives
P	=	paperclips
R	=	rope
S	=	salt and sugar
T	=	toothpicks
V	=	vest
W	=	watch
Z	=	zipper

Part IV

Braintempters

A History of the Ten-Square

A. Ross Eckler

Before the creation of the crossword puzzle in 1913, puzzle enthusiasts were constructing word squares—squares of letters in which every row (left to right) and every column (top to bottom) reads as an English word—of ever-increasing size: the six-square in 1859, the seven-square in 1877, the eight-square in 1884, and the nine-square in 1897. (Most appear in *The Key to Puzzledom*, published by the Eastern Puzzlers' League in 1906.)

But then progress stopped. Over the next eighty years, 869 eight-squares and 836 nine-squares were published in *The Enigma*, the monthly magazine of the National Puzzlers' league (successor to the Eastern Puzzlers' League in 1920). Most were laboriously constructed from the bottom up, aided by immense lists of eight-letter and nine-letter words in reverse-alphabetical order. Typically, such lists contained from 75,000 to 200,000 words drawn from many different sources and were issued in only a few copies. Puzzlers of that era apparently recognized that a ten-square was too difficult to construct by these methods; little attempt was made to assemble analogous lists of ten-letter words to aid in this endeavor.

In 1921, Paul M Bryan of the National Puzzlers' League constructed a special case of the ten-square: the tautonymic[1] ten-square, which appeared in the September *Enigma*.

For the past 33 years, **A. Ross Eckler**, has edited *Word Ways*, the only journal in the world devoted to wordplay.

[1] A "tautonym" is a scientific name in which the genus and species names are the same, such as rattus rattus (the black rat). Here "tautonymic" refers to an arbitrary word (ALALA) being similarly repeated.

```
G A P A S G A P A S
A L A L A A L A L A
P A R A N P A R A N
A L A N G A L A N G
S A N G A S A N G A
G A P A S G A P A S
A L A L A A L A L A
P A R A N P A R A N
A L A N G A L A N G
S A N G A S A N G A
```

Most of the words were drawn from place-names in the Philippine Atlas. Unfortunately, it contained one coinage—the duplication of ALALA, a Greek interjection mentioned in the Merriam-Webster New International Dictionary.

Arthur F. Holt, the constructor of the 1897 nine-square, believed himself to be the most talented wordsmith of his generation. (He was certainly one of the most successful winners of various contests, having won the first automobile offered as a prize, as well as a grand piano. He moved to Washington, D.C. shortly after the turn of the century; most believed that he wished to avail himself of the resources of the Library Congress in these endeavors.) He responded to Bryan's achievement with 22 tautonymic ten-squares published in the December *Enigma*. Unfortunately, he had to mine foreign-language works for a sufficient supply of tautonyms, and as a result his efforts were largely ignored. In 1946 Arthur L. Smith of the League praised a later Bryan tautonymic ten-square as the best of the lot.

```
A R A N T A R A N T
R E N G A R E N G A
A N D O L A N D O L
N G O S I N G O S I
T A L I N T A L I N
A R A N T A R A N T
R E N G A R E N G A
A N D O L A N D O L
N G O S I N G O S I
T A L I N T A L I N
```

The tautonymic ten-square briefly reappeared in 1973, when Dmitri Borgmann, another puzzler-for-revenue with a high opinion of his ability, generated two examples for the August and November issues of *Word Ways, the Journal of Recreational Linguistics*. Although these were based on more modern references, they were even less satisfactory than the earlier specimens. In one, the word ALGALALGAL appeared not twice but four times, and the other contained the non-dictionary phrases A SAIL! A SAIL! (in The Rime of the Ancient Mariner, by Coleridge) and RABBI, RABBI (from Matthew 23 of the New Testament).

Word researchers have always regarded the tautonymic ten-square as an unsatisfactory solution to the problem. However, no one seriously tried to construct a non-tautonymic example until Jeff Grant of Hastings, New Zealand presented three in the November 1985 *Word Ways*—the result of more than three years of work on his part. They relied on constructions such as SOL SPRINGS (individuals with the name Sol Spring, several of whom existed), SES TUNNELS (French for "his/hers/its tunnels"), IMPUTIEREN (German "to impute something") and BESSONNELS (a logically-formed diminutive of the French surname Besson). Nevertheless, his squares offered the first real hope that diligent research might result in a ten-square as acceptable as the nineteenth-century examples of smaller squares previously mentioned. Grant provided improved ten-squares in the November 1990 and November 1995 *Word Ways*, culminating with the following one in February 2002:

```
D I S T A L I S E D
I M P O L A R I T Y
S P I N A C I N E S
T O N Y N A D E R S
A L A N B R O W N E
L A C A R O L I N A
I R I D O L I N E S
S I N E W I N E S S
E T E R N N E S S E
D Y S S E A S S E S
```

DISTALISED: past participle of the verb distalise, to make more distal [Oxford English Dictionary Word and Language Service]. "This is especially true if buccal teeth have been distalised or retracted with or without headgear" [Are Extractions Necessary? http://www .orthotropics.com/professionals/dentists_areextractions.html]

IMPOLARITY: lack of polarity, particularly nonexpression of opposite emotional extremes, absence of mood swings—a rare technical term in behavioral psychology. "This battery of tests will emphasize those functions shown to be impaired on an empirical basis (problem solving, abstraction, linguistic ability) and theoretical basis (behavioral self-regulation, impolarity, shift of set, sustained effort, and attention)." [Ming T. Tsuang and Michael J. Lyons, An Identical

Twin High-Risk Study of Biobehavioral Vulnerability, http://www
.nida.nih.gov/pdf/monographs/monograph159/081-112_Tsuang.pdf]

SPINACINES: plural of spinacine, an amino acid derived from the much
commoner and more important amino acid histidine [Webster's Third].

TONY NADERS: persons named Tony Nader, such as residents of
Chatsworth, California and Torrington, Connecticut [Google,
www.whitepages.com].

ALAN BROWNE: a common name found in many telephone directories
throughout the world; the most well-known person with this name is
an American bank consultant [Who's Who in America, 45th edition,
1988-89].

LA CAROLINA: a town on the southern slope of the Sierra Morena
mountain range in Spain, 32 miles north of Jaen [Webster's New
Geographical Dictionary].

IRIDOLINES: plural of iridoline, an oily liquid compound derived from
coal tar [Funk & Wagnalls New Standard Dictionary, 1963].

SINEWINESS: the state or quality of being sinewy; firm strength, tough-
ness [Webster's Third].

ETERNNESSE: variant of eternness, a rare and obsolete synonym for
eternity [OED, 1608 quote].

DYSSEASSES: plural of dysseasse, a 16th-century spelling of the noun
"disease" [OED].

Finding word squares by hand is a hit-or-miss proposition; it is virtu-
ally certain that excellent squares contained in a given set of words will be
overlooked because of the astronomically large number of possible combinations. For 30 years or more, it has been obvious that computers offer a
solution to the problem, if they can be made sufficiently powerful to run
through all the possibilities in a reasonable time.

In 1975 M. Douglas McIlroy of Bernardsville, New Jersey searched by
computer for seven-squares constructible from the 9,663 seven-letter words
in the 7th edition of the Merriam-Webster Collegiate, finding 54 examples,
most of them new. The results appeared in the November *Word Ways*.
In 1988 Eric Albert of Auburndale, Massachusetts employed a far more
powerful computer to look for eight-squares and nine-squares. He soon
found 749 eight-squares in a 50,000-word list. However, the 63,000 nine-
letter words drawn from various dictionaries took a lot longer to process.

After running off and on for nearly a month, his computer finally disgorged a single square. (Two years later, his new computer could have done the full search in a weekend!) This square appeared in the October 1989 *Enigma* and the November 1991 *Word Ways*; all words, in fact, were in the Second Edition of the Merriam-Webster Unabridged.

But could the computer find a ten-square? In 1976 Frank Rubin of Wappingers Falls, New York programmed his computer to look for ten-squares among the 35,000 words in Webster's Second, and found one that contained eight words and two nonsense sequences (SCENOOTL and HYETNNHY); this appeared in the February 1977 *Word Ways*.

In a landmark paper in the February 1993 *Word Ways*, Chris Long devised a probability model to predict how large a word-list is needed to have a reasonable chance of generating a word square. Using Bayes' Theorem from probability theory, he showed that the expected number of n-squares found among a set of W n-letter words is

$$E = W^n (15.8)^{-n(n-1)/2}$$

under the assumption that the "words" in the list are formed independently and at random from a stockpile of letters whose frequencies match those in a typical word-list. The particular number 15.8 comes from an experimental measurement of the reciprocal of the squared probabilities of each letter (A–Z) in an English word-list. Setting $E = 1$, one can ascertain the size S of the word-list required, on the average, to yield a single square:

$$S = (15.8)^{(n-1)/2}.$$

For three-squares through ten-squares, the predicted support is, respectively, 16, 63, 250, 992, 3,944, 15,678, 62,320, and 247,718.

Of course, there is no guarantee that a real word list of size S will yield a word square, but this formula offers a reasonable way to extrapolate from known results for small word squares. If 9,663 seven-letter words yield 54 squares, then $(9,663)/54^{1/7} = 5,459$ words should yield one square; similarly, if 50,000 eight-letter words yield 749 squares, then 21,862 words should yield one square. These two quantities are 1.38 times as large as the theoretical values of 3,944 and 15,678. Applying this factor to the ten-square support value, a list size of $(1.38)(247,718) \approx 342,000$ is needed.

This number is far larger than any dictionary or combination of dictionaries can supply; it is evident that the ten-square must rely on other, larger databases. Is there any such database in computer-readable form?

If names of people such as JAMES SMITH or MARY MILLER are allowed, a new approach recently occurred to me. Using the website http://www.census.gov/genealogy/names, one can identify the 90 thousand commonest surnames in the United States, along with several thousand of the commonest male and female first names. The product yields far more than the 342,000 names required for a square, so one can afford to be choosy and use only the more frequently used names on these lists. Steve Root of Westboro, Massachusetts found, in fact, that 500,326 combinations generated 84 ten-squares; of these, 23 contained real people as evinced by their presence in the AOL White Pages, Social Security death records, or Internet searches. One of the best of these squares is given below (numbers at the right indicate the number of AOL White Pages listings, with + indicating ten or more):

```
L  E  O.  W  A  D  D  E  L  L    1
E  M  M  A.  N  E  E  L  E  Y    1
O  M  A  R.  G  A  L  V  A  N    5
W  A  R  R  E  N.  L  I  N  D    9
A  N  G  E  L.  H  A  N  N  A    2
D  E  A  N.  H  O  P  P  E  R    +
D  E  L  L  A.  P  O  O  L  E    3
E  L  V  I  N.  P  O  O  L  E    3
L  E  A  N  N.  E  L  L  I  S    3
L  Y  N  D  A.  R  E  E  S  E    5
```

The support is 325,000, a factor only 1.31 greater than the theoretical support of 247,718. All 23 of the squares are given in the May 2002 *Word Ways*.

Recently the US Board of Geographic Names has placed the National Imagery and Mapping Agency database on the Internet. Adding the ten-letter names from this list to 260,000 words drawn from the OED and 100 other references, Rex Gooch of Letchworth, England created a list of some 700,000 words and names. He estimates that it would take his computer some 39 years to search for all the possible ten-squares. Yet the scaling formula gives him reason for hope—if 324,000 words on the average generate one square, then 700,000 words should generate between two and three thousand (one square every few days). By June 2002 the best square found was the one in Figure 1.

The four locations with latitude and longitude are taken from NIMA, the National Imagery and Mapping Agency's database. Mahras Dagi is equivalent in form to Mount Everest. Tautologia is listed under the etymology for tautology in the OED. One can look forward with some confidence to the discovery of a ten-square of even higher quality in the near future.

All ten-squares in this article are symmetric, i.e., the same set of words appears both horizontally and vertically. Asymmetric ten-squares (all

A B A P T I S T U M   Webster's Second
B A H R A M T A P A   in Azerbaijan, 39°44ʹ latitude, 47°57ʹ longitude
A H L E R B R U C H   in Germany, 52°12ʹ latitude, 8°29ʹ longitude
P R E P A R A T O R   OED
T A R A D A N O V A   in Russia, 54°45ʹ latitude, 86°41ʹ longitude
I M B R A N G L E S   OED (imbrangle, verb)
S T R A N G F O R D   in Hereford and Worcester, England
T A U T O L O G I A   Late Latin (or Greek)
U P C O V E R I N G   OED
M A H R A S D A G I   in Turkey, 36°43ʹ latitude, 33°17ʹ longitude

Figure 1.

words different) are possible, but none have been found; it is estimated
that a word stock of 1,350,000 words is needed to have a reasonable chance
of producing one.

Configuration Games

Jeremiah Farrell, Martin Gardner, and Thomas Rodgers

Tripos is a two-person game whose play takes place on the colored 3×3 grid shown in Figure 1. One player uses pennies and the other player uses dimes. The players alternate placing one of their coins on an empty square. The game ends when one of the players (the winner) acquires three of their coins in a row, in a column, or on squares of one color.

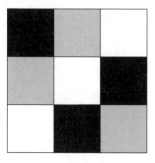

Figure 1. Tripos.

Jeremiah Farrell is a mathemagician, has Erdős number 2, and is a friend of Dr. Irving Joshua Matrix. **Martin Gardner** is the father of recreational mathematics, most famous for his 25-year Mathematical Games column in Scientific American. He has written more than 65 books throughout science, mathematics, philosophy, literature, and conjuring. **Thomas Rodgers** organizes the Gathering for Gardner.

Tripos is similar to tic-tac-toe, but there are major differences. For instance, there is no favored square in Tripos corresponding to the powerful central square in tic-tac-toe, since each square of Tripos is on exactly one row, exactly one column, and exactly one color (a generalized kind of "diagonal").

It seems unfair for the first player to have an extra fifth move, so we decree as a rule that each player has only four coins to put in play. In fact, we choose to be even more harsh: If the first player cannot win in four moves, then the win is given to the second player. Under these rules, Tripos can have no draws (as tic-tac-toe does). This means that one of the players has a winning strategy. Can you find it?

Pappus's Mousetrap is a game whose play takes place on the configuration of 9 points lying on 9 lines shown in Figure 2. As in Tripos, players take turns placing pennies and dimes until, in this case, one of them acquires three of their coins in a single line. For example, the first player would win by placing pennies on the points labelled 6, 4, and 3. As in Tripos, each player is allotted four coins, and the second player is given the win if the first player fails to win in four moves. Once again, there is a winning strategy for one of the players. Can you find it?

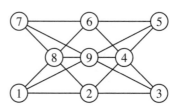

Figure 2. Pappus's Mousetrap.

Tri-Hex[1] is played like Pappus's Mousetrap, but on the configuration in Figure 3. Can you find which player has a winning strategy?

Finally, which two of these three games—Tripos, Pappus's Mousetrap, and Tri-Hex—are most alike?

It may seem as if Pappus's Mousetrap and Tri-Hex have the most in common. But surprisingly, the two that are most alike are Tripos and Pappus's Mousetrap: They are, in fact, the same game! If you label the squares of Tripos as in Figure 4, it's easy to check that a winning position in Tripos corresponds to a winning position in Pappus's Mousetrap and vice versa.

---

[1]O'Beirne commercialized Tri-Hex under this name, so we use it as the primary name. However, we also call it "O'Beirne's Mousetrap" by analogy to our names for other similar configurations.

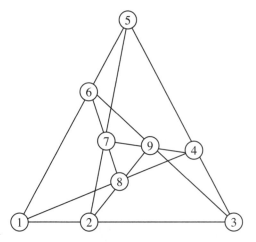

Figure 3. Tri-Hex.

5	7	6
9	8	4
1	2	3

Figure 4. Tripos, labeled to match Pappus's Mousetrap in Figure 2.

How do we know Tri-Hex isn't also the same as Tripos? The answer lies in the "misgraphs" of the games: the graphs formed by connecting points that do *not* line up with each other. For Tripos (a.k.a. Pappus's Mouse-trap), the misgraph consists of three triangles as shown in Figure 5(a). For Tri-Hex, the misgraph is a triangle and a hexagon as shown in Figure 5(b). The misgraphs also provide the key to a winning strategy for each game. For Tripos, the first player may as well pick the square labeled 1. If the second player starts with 4 or 7—the other two squares in 1's triangle in the misgraph—the first player can win by picking the remaining square in the triangle: The next choice by the second player will be blocked by the first player in a way that sets up two different ways for the first player to win on his next move. If, on the other hand, the second player opens by selecting a square in one of the other triangles of the misgraph, the first player can respond with a move that forces the second player to block with a move in the same triangle, leaving the first player free to make a move that sets up two ways of winning. Thus Tripos has a winning strategy for the first player, who may open with any move.

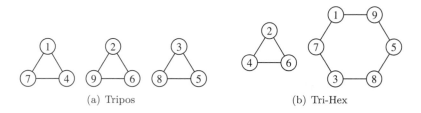

(a) Tripos   (b) Tri-Hex

Figure 5. Comparison of two misgraphs.

(When playing second against a novice, the first alternative in the analysis above—that is, picking in the same triangle of the misgraph—is a good move to make, since the first player's *only* winning move is to pick the last square in the triangle.)

Tri-Hex also has a winning strategy for the first player, but it *requires* opening with a move in the misgraph's triangle, not the hexagon. We leave the rest of the analysis to the reader.

These two games—Tripos and Tri-Hex—are examples of games based on what are called "symmetric configurations." See [4] for more details. More precisely, they are based on $(9, 3)$ configurations. In general, an (n, r) configuration is a collection of n "points" and n "lines" subject to the following requirements:

R1: Any two points belong to at most one line.

R2: Each line has r points, and each point belongs to r lines.

These requirements limit the number of distinct solutions (up to isomorphisms) for each type of configuration (n and r). If we restrict attention to $(n, 3)$ configurations, Gropp [4] reports only one solution for $n = 7$ and for $n = 8$, three for $n = 9$, and ten for $n = 10$. Figure 6 shows geometric realizations of these configurations. Note that for the $n = 7$ and $n = 8$ cases, one "line" is represented by a circle.

Each configuration defines a game. Fano's Pyramid, whose misgraph is totally disconnected, is not terribly interesting; it is an easy draw (that is, a win for the second player) as long as the second player blocks the first player's third move. Asteroid is slightly more interesting. The misgraph has four pieces: 1–8, 2–5, 3–7, and 4–6. The first player can win by starting on, say, 1. If the second player picks 8, then the first player can win by picking any other number; on the other hand, if the second player picks any

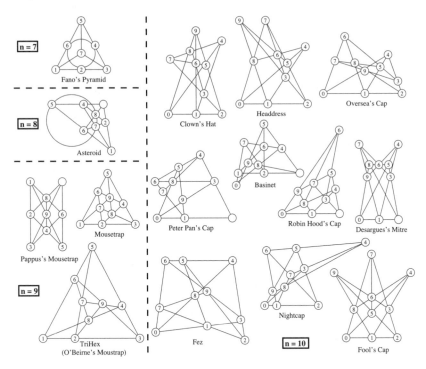

Figure 6. All $(n, 3)$ configurations for $n = 7, 8, 9, 10$, as enumerated in [4].

other number, then the first player can set up a win by picking 8. (What if the first player starts on 2, 5, 3, or 7?)

The misgraph for Mousetrap is a single cycle: 1–4–7–5–2–9–3–6–8–1. The cyclic relabelings of the configuration shown in Figure 7 show that all starting points are essentially the same. (In the jargon of group theory, the automorphism group of the configuration acts transitively on the configuration.) So the first player may as well start by picking 1. It turns out that the second player has a unique winning move. Can you find it?

We leave it to the interested reader to analyze the $(10, 3)$ configuration games, but we close with one final question. The game Pentacles [7] is played on the $(10, 3)$ configuration in Figure 8 with the same rules as Tripos and Tri-Hex.[2] However, the Pentacles configuration doesn't look like any of the $(10, 3)$ configurations in Figure 6, but it must be equivalent to one of them. Which one?

For related word puzzles based on symmetric configurations, see [1, 2].

---

[2]Figure 8 is used in a different context by O'Beirne in [6], so we also call the game "O'Beirne's Stars". Other configurations of O'Beirne are described by Gardner [3].

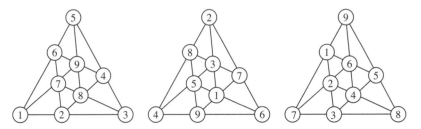

Figure 7. Relabeling Mousetrap according to the misgraph cycle: as we proceed left to right, we replace 1 with 4, 4 with 7, 7 with 5, etc. All games we obtain are equivalent.

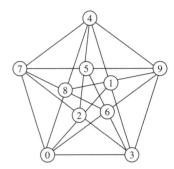

Figure 8. Pentacles.

Figure 9. We have adapted Pentacles and Tripos so that they can be played effectively by both blind and sighted persons. They have proved popular with students at the Indiana School for the Blind. In Pentacles, the edges are represented by dowels. In Tripos, the 3×3 grid positions are represented by raised diamond, heart, or serrated circle shapes. (See Color Plate VII.)

Solutions

Mousetrap: The second player's unique winning move is 7.

Pentacles: Pentacles is equivalent to Robin Hood's Cap, using the labelings in Figures 6 and 8.

References

[1] Jeremiah Farrell. Games on word configurations. *Word Ways: The Journal of Recreational Linguistics*, 27(4):195–205, November 1994.

[2] Jeremiah Farrell. Puzzles and games on word configurations. *Word Ways: The Journal of Recreational Linguistics*, 34(3):243–249, November 2001.

[3] Martin Gardner. *Mathematical Magic Show.* New York: Knopf, 1977, page 69.

[4] Harald Gropp. Configurations. In *CRC Handbook of Combinatorial Designs*, Boca Raton, FL: CRC Press, 1996.

[5] Jeremy Thomas Lanman. The Taxonomy of Various Combinatoric and Geometric Configurations. Undergraduate Honors Thesis, Butler University, May 2001.

[6] T. H. O'Beirne. *Puzzles and Paradoxes.* New York: Dover, 1984, page 109.

[7] Thomas Rodgers. Pentacles. Handout at G4G5, April 2002.

BlackJack Stacks

Harold Cataquet

A stacked deck is an arrangement of cards used by a magician or gambler. The magician uses a stacked deck to achieve a magical effect, or give the impression of superior card handling ability. The gambler uses a stacked deck to win money from unsuspecting players. Over the years, magicians and gamblers have designed hundreds of excellent poker stacks, but the subject of BlackJack stacks has been largely ignored.[1]

In the magic literature, there are three BlackJack stacks:

- Paul Studham in 1942 [4]

- Monte Cooper in 1961 [1, 5, 6]

- Eddie Fields [3]

However, this list is deceptive. The Studham stack is not very effective[2] and the Fields stacks is not a genuine BlackJack stack as the spectator never cuts the deck and, in performance, the role of dealer alternates between the

---

Harold Cataquet was born and raised in New York, but now lives in England. He runs his own financial training consultancy, and designed one of the most advanced trading simulators. He is also an award winning magician and designer of puzzles.

[1]In the gambling literature, BlackJack stacks probably outnumber poker stacks. See, for example, L. Miller "BlackJack Exposed" 1970.

[2]On average, in playing out the deck, the player wins 5 hands, and the dealer wins 4 hands (all draws).

magician and the spectator in a predetermined sequence. So, the Cooper stack is probably the *only* BlackJack stack in the magic literature.

The Cooper stack is A532670080094. That is, the deck is arranged in four groups of 13 cards (the suits are unimportant) with the tens (ten, Jack, Queen, and King) occupying any of the 0 slots. The Ace is the top card of the face down deck, the 5 is the second, the 3 is the third, etc. Note that the order doesn't have to begin with an Ace. For example, you can use 2670080094A53. However, each block of thirteen must have the identical order. So, one complete deck order might be:

$$80094A532670080094A532670080094A532670080094A5326700$$

The beauty of this stack is that the deck can be freely cut without affecting the performance of the stack.

Unfortunately, the Cooper stack is not as strong as originally claimed. In particular, Cooper and others have claimed that the dealer can never lose more than two hands in playing out the deck. In truth, the dealer can lose all eight hands.[3]

To see this, suppose that the spectator cuts the deck so that the order from the top was 080094A532670. In the first hand, the player gets 20, and the dealer gets 18, so the player wins (1–0 to the player). In the next hand, the player gets a soft 20 and the dealer gets 9. Now, suppose the player decides to take another card. Now he has 13 (with the 3), so he decides to take another card. Now he has 15 (with the 2), and he again decides to take another card. Having 21 (with the 6), the player sticks. The dealer takes the next card (7) which gives him 16. He is thus forced to take the next card, which is a 10, so he busts. The player is now 2 games up, but, more importantly, this sequence has exhausted exactly 13 cards. So, the next thirteen cards will be exactly the same. If the spectator is smart enough to remember the pattern, the dealer will lose 8 games in a row, and exhaust the deck.

Moreover, a worst-case analysis shows that it is possible for the player to "steer" the Cooper stack into this vicious cycle from any initial cut, suffering at most two losses along the way. For example, suppose the cut is 094A532670080. In the first hand, the player gets 14 and the dealer gets a soft 20. The player stands, and loses the hand. Next, the player gets 7 (5 and 2) and the dealer gets 9 (3 and 6). Counterintuitively, the player stands, causing the dealer to hit twice (getting the 7 and the 10) and bust. From then on the player can remain in the vicious cycle between 0800... and 94A5... as described above, and win all the remaining games.

---

[3]When playing out a complete deck, the number of hands will range from 8 to 12.

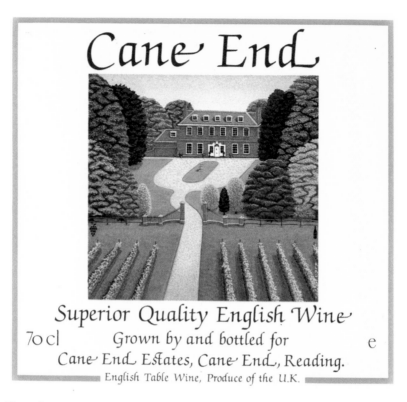

Cane End

Superior Quality English Wine

7o cl   Grown by and bottled for   e
Cane End Estates, Cane End, Reading.

English Table Wine, Produce of the U.K.

Plate I. (See Page xvi.) Follow the dark lines from the grape icon in the middle left, through the bushes, across the lower ledge of the upper roof, through the bushes again, to the bottle icon in the middle right.

Plate II. (See Page 17.) Zodiac figural puzzles.

Plate III. (See Page 5.) Justice Cup with base. Late 19th / early 20th century.

Plate IV. (See Page 6.) Justice Cup showing the drain holes in the base of the cup and the top of the base.

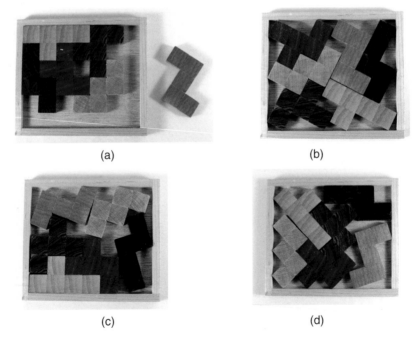

(a)

(b)

(c)

(d)

Plate V. (See Page 221.) Stewart Coffin's design #175. (a) Trying a "normal" solution; (b) The elegant solution; (c-d) Two not-so-elegant solutions.

Plate VI. (See Page 81.) Absolute Martin.

Plate VII. (See Page 98.) Pentacles and Tripos adapted so that they can be played effectively by both blind and sighted persons. In Pentacles, the edges are represented by dowels. In Tripos, the 3 × 3 grid positions are represented by raised diamond, heart, or serrated circle shapes.

Plate VIII. (See Page 16.) Carved and decorated wooden Turtle Puzzle.

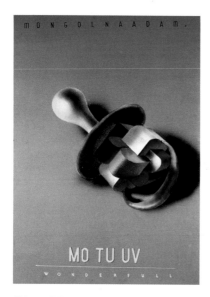

Plate IX. (See Page 13.) Puzzle art by Tumen-Ulzii.

Plate X. (See Page 19.) Turtle chess set.

So, is there a better stack? Well, there are at least 50. In this paper, I offer a few stacks which are substantially better than the Cooper stack.

There are $12!/4! = 19{,}958{,}400$ different 13-card stacks. Each of these stacks has 13 different starting positions. Over the past year, all of these stacks and all of the possible starting positions have been analyzed using a number of playing rules, and playing strategies. For simplicity, the initial analysis assumed that the player could see the next card, and was always trying to win. We also assumed that the dealer wins on all ties; five-card Charlie[4] is not an automatic win for the player; and no splits on tens are allowed.[5]

In all versions of BlackJack, the first card is dealt to the player, the second to the dealer, and the third to the player. All the cards are dealt face up. Under American rules, the fourth card is dealt face down to the dealer; under European rules, the dealer does not take a card until the player has exhausted his turn. Under both sets of rules, the dealer must hit on 16 or less, as well as "soft" 17 (Ace and a six). Both sets of rules also consider a two-card 21 to be an automatic win for the player (i.e., the dealer does not get a chance to tie him).

While the Cooper stack is the best known American stack, there is also one European stack in the magic literature. Attributed to Fred Robinson, the stack is A637054800029; it appears in [2]. Unfortunately, there is no record of the rules that Robinson used (e.g., the dealer may have had complete freedom as to how to play his hand). However, using the standard rules for European BlackJack, the stack performs only slightly better than the Studham stack does for American rules.

Winning Stacks for the Dealer

For American rules, the best stack is A576020090483. There are only two hands the player can win on: A7 and 8A. If the player holds A7 (soft 18), the dealer shows 5 with 6 in the hole. The rule of thumb in BlackJack is to stick on soft 18, in which case the dealer would draw a 10 for the win. But if instead the player takes a hit, he gets the 10 (for hard 19), and the dealer then draws a 2 and then a 10, and busts. In the 8A (soft 19) case, the player can win by taking a hit, which gives him the 7, for a total of 16;

---

[4]In many casual games, if a player gets five cards without going over 21, he automatically wins. This is never seen in casinos.

[5]Technically, no splits are allowed. However, with the setup described and a simple cut, the player will never be dealt a pair, so we can allow these. What the player may see is two tens (e.g., a King and a Jack), so we don't allow these to be split.

the dealer, shows 3 with a 5 in the hole, draws a 6 for 14 and then a 10 to bust.

Under European rules, seven stacks performed well. They are

A760839050240	(68, 30, 52, 4A)
A760839040250	(68, 30, 42, 5A)
A860930025074	(69, 30, 02, 57)
A523978000460	(04, 40, 6A, 05)
A869030025074	(60, 30, 02, 57, 48)
A740859030260	(48, 50, 32, 6A, 07)
A352087900640	(A5, 32, 06, 60, 4A),

where the items in parentheses are the hands the player can win if he plays his cards right. For example, in the first three stacks, the player can win the 68 hand by sticking on 14, and forcing the dealer, who shows a 10, to draw a 3 and then bust on the next card, which is a 9 for the first two stacks and a 10 for the third. In a real game of BlackJack, a player would rarely stick on 14 against a dealer's 10. However, performed as a magic trick, the dealer could probably persuade the player to take a card by showing him it was a 3 and would unambiguously improve the player's total.

If we change the rule about two-card 21 being an automatic win for the player, then A760839050240 (European) continues to be the best European stack, while A576020090483 (American) goes from being the best stack to number five in the list. The European stack is therefore quite robust to rule changes. However, when presenting this as a magic trick, the dealer does need to be aware of all of the losing hands, and be prepared to play an unnatural role (encouraging the player to take another card).

If two-card 21 is no longer an automatic win, then the best American stack is A080492057603. With this stack, the player has only one way to win. Specifically, if the player gets BlackJack (0A), he will lose on a push as the dealer also gets 21 (308) albeit with three cards. But, if the player hits on BlackJack (which goes against all logic), the player gets 19 (0A8), and the dealer subsequently loses with 23 (300). From a gambling standpoint, there is something to be said for the counterpunch given by the dealer quashing a two-card 21. However, with a player who is very familiar with casino rules, you need to emphasize at the outset that a two-card BlackJack is *not* an automatic win.

Even though all of these stacks are the "best" in one form or another, I am partial to A405097068302, even though there are two potential losing hands. In this stack, the player always wins when he has a two-card 21, and that will not occur more than once in a deck. But, what I like about this stack is that more hands are played out than in the other stacks. So, if you are playing for money, the player will lose one or two hands more than the "better" stacks mentioned above. In addition, this new stack gives the dealer the most outright wins (as opposed to the fewest losses which is how the "best" stacks were defined). So, although A080492057603 is the best stack, A405097068302 is the stack that I use most often in my performances.[6]

Winning Stacks for the Player

There are no perfect BlackJack stacks for the player, because the player has various ways of playing each hand, whereas the dealer's actions are always automatic. Nonetheless, there are stacks in which the *player* will win every hand. These stacks should be used when the spectator is to play the role of the dealer and the magician will be the active player. Under American rules, A460352000987 and A982600730540 enable the player to always win. Here, for example, is one way to guarantee wins for A460352000987:

A6 vs. 40	Player hits and gets 20 (A63);	dealer gets 19 (405)
40 vs. 63	Player sticks with 14 (40);	dealer busts (63520)
63 vs. 05	Player hits twice and gets 21 (6320);	dealer busts (050)
05 vs. 32	Player sticks at 15 (05);	dealer busts (3200)
32 vs. 50	Player hits and gets 15 (320);	dealer busts (500)
50 vs. 20	Player sticks at 15 (50);	dealer busts (200)
20 vs. 00	Player hits and gets 21 (209);	dealer has 20 (00)
00 vs. 09	Player sticks at 20 (00);	dealer has 19 (09)
09 vs. 08	Player sticks at 19 (09);	dealer has 18 (08)
08 vs. 97	Player sticks at 18 (08);	dealer gets 17 (97A)
97 vs. 8A	Player hits and gets 20 (974);	dealer has 19 (8A)
8A vs. 74	Player sticks at 19 (8A);	dealer gets 17 (746)
74 vs. A6	Player hits and gets 21 (740);	dealer gets 17 (A6).

Note that since this stack never deals the player a two-card 21, the rule about it being an automatic win is irrelevant.

Under European rules, one player-friendly stack is A060495823070. Another nice stack for European rules is A070305024869. This latter stack requires an extra word of caution, because the player will lose the first hand if an Ace is cut to the bottom: 00 vs. 7 loses on a draw (the dealer gets

---

[6]In my lecture notes, which predate the current article, I called this the Cataquet stack. Looking at the stack, the two losing hands are:

Player has 21 (A0) and the dealer is showing a 4 (5 in the hole). The player sticks, and the dealer then gets a 10, losing with 19 against the player's BlackJack.

Player has 21 (A0) and the dealer showing a 4 (5 in the hole). The player hits and gets a 10, for a total of 21. The dealer then gets a 9, losing with 18 against the player's 21.

20 with 730). However, this can only happen on the very first hand (the strategy for playing subsequent hands avoids cycling back to it), and if the magician notices that the Ace is on the bottom of the deck, he can ask the dealer to cut the deck, thereby preventing the draw, or else he can play the first hand as a come-on before the actual betting begins.

Presentational Tips for Magicians

If you are going to do a BlackJack presentation, the key to a successful presentation is having the spectator believe that the deck was thoroughly shuffled and freely cut before the cards were dealt. Therefore, it is important that the magician give the deck one or two false shuffles, and then let the spectator have a free cut of the deck. Even better, begin with the stacked deck in your pocket. Now perform a few card tricks using a second matching deck. When you are finished, put the deck in your pocket and then, as if suddenly remembering that you wanted to show them one more effect, reach into your pocket and remove the stacked deck.

I also have the following suggestions:

- Set up the deck and crimp the 13th, 26th, and 39th card in the deck (these all have the same value as the 52nd card). Cut the deck below the first crimp and table the packet. Repeat until you have four piles on the table. Now, let the player reconstruct the deck by picking up the four stacks in any order.[7] Now, let him give the deck a complete cut. What could be fairer than that?!

- If you want, you can let the player see the next card. Don't do it straight away. Instead, wait until the player has lost a few hands, and then offer him some help and thereafter show him the next card. Similarly, you can do the same and show him the dealer's cards. This builds on the impossibility of beating the magician.

- You could play the entire deck face up from the beginning. Just reverse the order of the stack (i.e., the top card of the face down deck becomes the top card of the face up deck). Now, from the start, the player can see the next card.

Throughout the presentation, it is important that the magician point out that if the player had just cut one more card, or one less card, he would

---

[7]It should be clear that all four tabled piles have exactly the same order, differing only in suit composition.

have been receiving the cards that the dealer has been getting (and winning with).

Acknowledgments

Thanks to Paul Burrowes for his computer analysis and comments on earlier drafts of this paper. Thanks also to Tomas Blomberg for suggesting the idea of looking at all the possible player hands. Previously, I had only looked at "intelligent play." This alternative approach identifies all the ways that the player can win, but ignores how often the situation can occur in actual play. For example, the Cooper stack has eight possible losing hands, but two of those can occur four times in playing out a single deck. So, it is possible for the dealer to lose all eight hands. However, the real benefit of the Blomberg approach is to pinpoint the instances where the dealer can lose and what his response should be. For example, suppose that the player has 19 and the dealer has 12, and that the next 2 cards in the deck are 2 and 9. If the player sticks at 19, then the dealer busts. But, if the dealer shows the next card and encourages the player to take it (which would not require much effort), then the dealer wins on a draw.

References

[1] Monte Cooper, "Blackjack Runup." *Hugard's MAGIC Monthly*, September and October 1961, p. 23.

[2] Peter Duffie, *Virtual Miracles*, 2000. World Wide Web. <http://www.peterduffie.pwp.blueyonder.co.uk/bookspp.htm>.

[3] Jon Racherbaumer, "BlackJacked!" In *In a Class by Himself: The Legacy of Don Alan*, L&L Publishing, Tahoma, CA, 1996, pp. 181–186.

[4] Paul Studham, "Black Jack Magic." In *Ireland's 1942 Year Book*, Ireland Magic Company, Chicago, 1942, pp. 123–124.

[5] Nick Trost, "The BlackJack Demonstration." In *Expert Gambling Tricks with Cards*, 1976, pp. 23–24.

[6] Allan Slaight "Unkindest Cut of All." *Apocalypse* 2:4, April 1979, p. 186.

Five Algorithmic Puzzles

Peter Winkler

Puzzles with Parameters

Many fascinating mathematical puzzles revolve around algorithms. Typically, you (the victim) are presented with a 'situation' together with a collection of possible operations, and a target state. You may or may not be able to exercise choice in applying the operations. You are asked: Can you reach the target state? Or perhaps: Can you *avoid* reaching the target state? And sometimes: In how many operations?

One approach to such problems is to find a parameter P—some kind of numerical rating of states—which somehow encapsulates progress toward the target. Consider, for example, this practice problem from an old Russian Olympiad:

> Suppose that you are given an $m \times n$ array of numbers and permitted at any time to reverse the signs of all the numbers in any row or column. Prove that you can arrange matters so that all the row sums and column sums are non-negative.

Peter Winkler is Director of Fundamental Mathematics Research at Bell Labs, Lucent Technologies. He is the author of *Mathematical Puzzles: A Connoisseur's Collection*, recently published by A K Peters, Ltd.

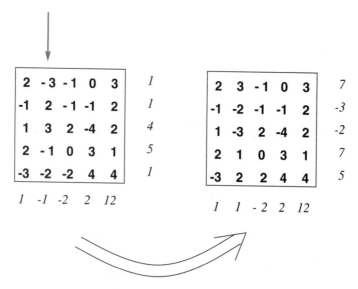

Figure 1. Flipping a column.

Of course, flipping a row that has a negative sum will fix that sum but possibly ruin some column sums. How can you be sure to make progress?

Perhaps the most obvious parameter P, the number of lines (rows and columns) with non-negative sum, is the wrong parameter because it could go down even when a line with negative sum is flipped. Instead, let's try setting P equal to the sum of all the entries in the array. Flipping a row with sum $-s$ increases P by $2s$ because P can be written as the sum of all the row sums (and similarly for columns). Because there are only finitely many reachable situations (actually, no more than 2^{m+n}), and P goes up every time you flip a negative-sum line, you must reach a time when all the line sums are non-negative.

In general, mathematical puzzles about reachability of a target state via specified operations ask you to prove a statement of one of the following types (though you are not necessarily told which):

1. There is a (finite) sequence of operations that reaches the target state;

2. Any sequence of operations will eventually reach the target;

3. Every sequence of operations reaches the target in the same number of steps;

4. No sequence of operations can reach the target.

Typically the operation changes some aspect of the situation for the better, while possibly losing ground elsewhere. How can you determine whether the target is reachable?

For problems of Type 1, you want to show that until the target is reached, there is always an operation (or sequence of operations) available that improves P. To make sure that you don't get caught in Zeno's paradox (making smaller and smaller steps, and never reaching the target value), you may have to show that P can always be improved by at least a certain amount, or that there are only finitely many possible situations.

For problems of Type 2, you do the same except that now you show that *every* choice of operation improves P.

For problems of Type 3, you show that every operation improves P by the same amount.

For problems of Type 4, you show that *no* operation improves P, yet attaining the target requires improvement.

The Russian Olympiad problem was a Type 1 problem, but as you see it could also have been phrased as a Type 2 problem by specifying that only negative-sum lines may be flipped, then asking you to show that you *will* reach a point when all the line-sums are non-negative.

For the problems below, considerably more imagination may be required to find a parameter P that works.

Five Problems

The Infected Checkerboard

An infection spreads among the squares of an $n \times n$ checkerboard in the following manner: if a square has two or more infected neighbors, then it becomes infected itself. (Neighbors are orthogonal only, so each square has at most four neighbors.)

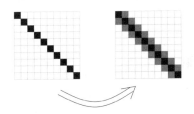

Figure 2. Infecting the checkerboard from the main diagonal.

For example, suppose that we begin with the n squares on the main diagonal infected. Then the infection will spread to neighboring diagonals and eventually to the whole board. Prove that you *cannot* infect the whole board if you begin with fewer than n infected squares.

Emptying a Bucket

You are presented with three large buckets each containing an integral number of ounces of some non-evaporating fluid. At any time you may double the contents of one bucket by pouring into it from a fuller one; in other words, you may pour from a bucket containing x ounces into one containing $y \leq x$ ounces until the latter contains $2y$ (and thus the former $x - y$).

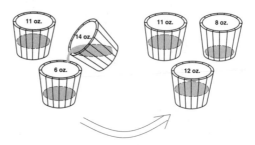

Figure 3. One step toward emptying a bucket.

Prove that no matter what the initial contents, you can, eventually, empty one of the buckets.

Pegs on the Half-Plane

Each grid point on the XY plane on or below the X-axis is occupied by a peg. At any time a peg can be made to jump over a neighbor peg (horizontal, vertical, or diagonal) and onto the next grid point in line, provided that point was unoccupied, after which the neighbor is removed.

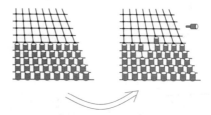

Figure 4. A diagonal jump puts a peg on the line $y = 1$.

Can you get a peg arbitrarily far above the X-axis?

Flipping the Polygon

The vertices of a polygon are labeled with numbers, the sum of which is positive. At any time you may change the sign of a negative label, but then the new value is subtracted from both neighbors' values so as to maintain the same sum.

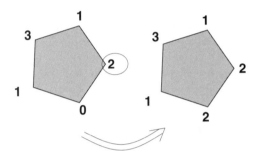

Figure 5. One flip on a pentagon.

Prove that inevitably, no matter which labels are flipped, the process will terminate after finitely many flips, with all values non-negative.

Breaking the Chocolate Bar

You have a rectangular chocolate bar marked into $m \times n$ squares, and you wish to break up the bar into its constituent squares. At each step you may break one piece along any of its marked vertical or horizontal lines.

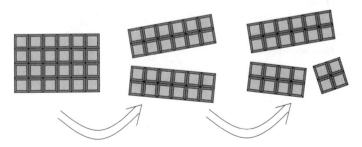

Figure 6. Two steps toward breaking up a chocolate bar.

Prove that every method finishes in the same number of steps.

Solutions

The Infected Checkerboard

This lovely problem appeared in the Soviet magazine KVANT around 1986, then migrated to Hungary; a probabilistic version is under study by mathematicians Gabor Pete and Jozsef Balogh. The problem reached me through Joel Spencer of the Courant Institute, who claimed there was a "one-word proof"! As you will see, this is only a mild exaggeration.

Would-be solvers, misled by the diagonal example, often try to show that there must be an initially infected square in each row or column; but that is far from true. Note for example that the configuration of sick squares shown in Fig. 7 spreads to the whole board.

Figure 7. Another way to infect the whole board.

Indeed there are myriad ways to infect the whole board with n sick squares, but apparently no way to do it with fewer. Some magic parameter P is needed here, but what?

The parameter is the perimeter! When a square is infected at least two of its boundary edges are absorbed into the interior of the infected area, and at most two added to the boundary of the infected area. Hence the perimeter of the infected area cannot increase. Because the perimeter of the whole board is $4n$ (assuming unit-length edges), the initial infected area must have contained at least n squares.

An additional exercise for those interested: prove that n initial sick squares are necessary even when the top and bottom of the board are joined to form a cylinder. If the sides are joined as well, forming a torus, then $n-1$ initial sick squares are sufficient (and necessary). The perimeter no longer works but another approach, found by Bruce Richter at the University of Waterloo and me, does the trick.

Emptying a Bucket

Yet another beauty from the former Soviet Union, this problem appeared in the fifth U.S.S.R. Mathematics Competition, Riga, 1971. The problem reached me via Christian Borgs, of Microsoft Research. I will give two solutions: a combinatorial one of my own, and an elegant number-theoretic one found by Svante Janson of Uppsala University, Sweden. I do not know which, if either, solution was the intended one.

In both solutions, P is the content of a particular bucket. In Svante's solution, we show how P can always be reduced until it is zero. In my solution, on the other hand, we show that P can always be *increased* until one of the *other* buckets is empty!

To do this, we first note that we can assume that there is exactly one bucket containing an odd number of ounces of fluid; for, if there are none, we can scale down by a power of 2, otherwise one step with two odd buckets will reduce their number to one or none.

Second, note that with an odd and an even bucket we can always do a reverse step, i.e., get half the contents of the even bucket into the odd one, by repeatedly pouring from the event bucket into the odd bucket. This is because each state can be reached from at most one state, so if you take enough steps you must cycle back to your original state; the state *just before* you return is the result of your "reverse step."

Finally, we argue that as long as there is no empty bucket, the odd bucket's contents can always be increased. For, if there is a bucket whose contents are divisible by 4, we can empty half of it into the odd bucket; if not, one forward operation between the even buckets will create such a bucket.

Here is Svante's solution, in his own words:

"Label the buckets A, B, and C with, initially, a, b, and c ounces of fluid, where $0 < a \leq b \leq c$. I will describe a sequence of moves leading to a state where the minimum of the three amounts is smaller than a. If this minimum is zero, we are home; otherwise we relabel and repeat.

Let $b = qa + r$, where $0 \leq r < a$ and $q \geq 1$ is an integer. Write q in binary form:

$$q = q_0 + 2q_1 + \cdots + 2^n q_n$$

where each q_i is 0 or 1 and $q_n = 1$.

Do $n + 1$ moves, numbered $0, \ldots, n$, as follows: in move i we pour from B into A if $q_i = 1$ and from C into A if $q_i = 0$. Because we always pour into A, its content is doubled each time so A contains $2^i a$ before the ith move. Hence the total amount poured from B equals qa, so at the end there remains $b - qa = r < a$ in B. Finally, observe that the total amount poured from C is at most

$$\sum_{i=0}^{n-1} 2^i a < 2^n a \le qa \le b \le c,$$

so there will always be enough fluid in C (and in B) to do these moves."

As far as I know, no one knows even approximately how many steps are required for this problem (in whatever is the worst starting state involving a total of n ounces of fluid). My solution shows that order n^2 steps suffices, but Svante's does better, bounding the number by a constant times $n \log n$. The real answer might be still smaller.

Pegs on the Half-Plane

This is a variation of a problem described in *Winning Ways for Your Mathematical Plays*, Vol. 2, by Berlekamp, Conway, and Guy (Academic Press, 1982)—see p. 715, "The Solitaire Army." We believe the problem was invented originally by the second author, John H. Conway of Princeton University. In Conway's problem diagonal jumps were not permitted; one can nonetheless get a peg to the line $y = 4$ without much difficulty, but an argument like the one below shows that no higher position can be reached.

With or without diagonal jumps, the difficulty is that as pegs rise higher, grid points beneath them are denuded. What is needed is a parameter P that is rewarded by highly placed pegs but compensatingly punished for holes left behind. A natural choice would be a sum over all pegs of some function of the pegs' positions. Because there are infinitely many pegs, we must be careful to ensure that the sum converges.

We could, for example, assign value r^y to a peg on $(0, y)$, where r is some real number greater than 1, so that the values of the pegs on the lower Y-axis sum to the finite number $\sum_{y=-\infty}^{0} r^y = r/(r-1)$. Values on adjacent columns will have to be cut, though, to keep the sum over the whole plane finite; if we cut by a factor of r for each step away from the Y-axis, we get a weight of $r^{y-|x|}$ for the peg at (x, y) and a total weight of

$$\frac{r}{r-1} + \frac{1}{r-1} + \frac{1}{r-1} + \frac{1}{r(r-1)} + \frac{1}{r(r-1)} + \cdots$$

$$= \frac{r^2 + r}{(r-1)^2} < \infty$$

for the initial position.

If a jump is executed, then at best (when the jump is diagonally upward and toward the Y-axis) the gain to P is vr^4 and the loss $v + vr^2$, where

v is the previous value of the jumping peg. As long as r is at most the square root of the "golden ratio" $\theta = (1 + \sqrt{5})/2 \approx 1.618$, which satisfies $\theta^2 = \theta + 1$, the net change $v(r^4 - r^3 - 1)$ can never be positive.

If we go ahead and assign $r = \sqrt{\theta}$, then the initial value of P works out to about 39.0576; but the value of a peg at the point $(0, 16)$ is $\theta^8 \approx 46.9788$ *by itself*. Because we cannot increase P, it follows that we cannot get a peg to the point $(0, 16)$. But if we could get a peg to *any* point on or above the line $y = 16$, then we could get one to $(0, 16)$ by stopping when some peg reaches a point $(x, 16)$, then redoing the whole algorithm shifted left or right by $|x|$.

We do not know the highest value of y for which the point $(0, y)$ is reachable, allowing diagonal jumps. Perhaps an industrious reader can close this gap.

Flipping the Polygon

This problem generalizes one which appeared in the International Mathematics Olympiad (submitted by a composer from East Germany, I am told) and subsequently termed "the pentagon problem."

The problem has many solutions, and can even be generalized further from n-gons to arbitrary connected graphs. However, the solution below stands out for its combination of elegance and strong conclusion. It was devised independently by at least two individuals, of whom one is Bernard Chazelle, Professor of Computer Science at Princeton University.

Let $x(0), \ldots, x(n-1)$ be the labels, summing to $s > 0$, and define the doubly infinite sequence $b(\cdot)$ by $b(0) = 0$ and $b(i) = b(i-1) + x(i \bmod n)$. The sequence $b(\cdot)$ is not periodic, but periodically ascending: $b(i+n) = b(i) + s$.

If $x(i)$ is negative, $b(i) < b(i-1)$ and flipping $x(i)$ has the effect of switching $b(i)$ with $b(i-1)$ so that they are now in ascending order. It does the same for all pairs $b(j)$, $b(j-1)$ shifted from these by multiples of n. Thus, flipping labels amounts to sorting $b(\cdot)$ by adjacent transpositions!

To track the progress of this sorting process, we need a parameter P that measures the degree to which $b(\cdot)$ is out of order, but is still finite. To do this let i^+ be the number of indices $j > i$ for which $b(j) < b(i)$, and i^- the number of indices $j < i$ for which $b(j) > b(i)$. Note that i^+ and i^- are finite and depend only on $i \bmod n$. Observe that $\sum_{i=0}^{n-1} i^+ = \sum_{i=0}^{n-1} i^-$; we let this sum be our magic parameter P.

When $x(i+1))$ is flipped, i^+ decreases by one and every other j^+ is unchanged, so P goes down by *exactly one*. When P hits 0, the sequence is fully sorted so all labels are non-negative and the process terminates.

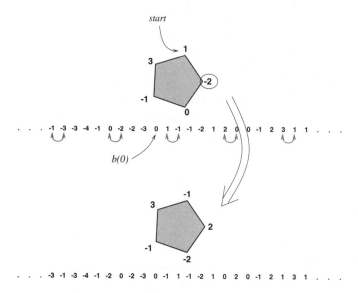

Figure 8. Flipping on a pentagon, while sorting a sequence.

We have shown more than asked: the process terminates in exactly the same number (P) of steps regardless of choices, and moreover, the final configuration is independent of choices as well! The reason is that there is only one sorted version of $b(\cdot)$; entry $b(i)$ from the original sequence must wind up in position $i + i^+ - i^-$ when the sorting is complete.

Breaking the Chocolate Bar

This one has been known to stump some *very* high-powered mathematicians for as much as a full day, until the light finally dawns amid groans and beatings of the head against the wall. We wouldn't want to deprive you, dear reader, of the opportunity to try it for yourself.

Acknowledgments

Inspiration for collecting these puzzles and much more comes from Martin Gardner's famous *Mathematical Games* column in *Scientific American*. Thanks also to my (former) friends who served as test victims for countless puzzles.

Polyform Patterns

Ed Pegg, Jr.

Polyforms are generalizations of Solomon Golomb's polyominoes in which squares are replaced by other kinds of polygons. Polyiamonds, for example, are based on equilateral triangles. Just as the number of polyominoes of size 1, 2, 3, etc. gives the sequence 1, 1, 2, 5, 12, 35, 108, etc., the number of polyiamonds gives the sequence 1, 1, 1, 3, 4, 12, 24, 66, 160, 448, etc. [18]. Polydrafters—polyforms made from 30-60-90 triangles (i.e., halves of an equilateral triangle)—are another example, with the sequence 1, 6, 14, 64, 237, 1,024, etc. The 14 tridrafters are shown below, arranged in the form of a trapezoid, with dotted lines outlining the individual 30-60-90 triangles.

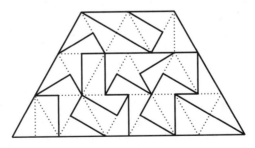

A former programmer for NORAD, **Ed Pegg Jr.** is the webmaster of mathpuzzle.com, and writes the weekly Math Games column for maa.org. He maintains library.wolfram.com for Wolfram Research.

Note that we are only counting the number of distinct final shapes, not the precise way they are formed. In particular, many polydrafters include a rectangle that can be cut into two 30-60-90 triangles in two different ways (along either diagonal).

The study of polyforms got a boost in early 1999, when Scottish lord Christopher Monckton introduced a new polyform puzzle, Eternity, with a million pound prize for the first solver. A former member of Margaret Thatcher's staff, Monckton designed Eternity during a bout of ill health. The 209-piece puzzle was so hard, it was thought, no one would be able to solve it.

A study of the publicity photo revealed the Eternity pieces are dodeca-drafters—that is, composed of twelve 30-60-90 triangles. The top poly-formists, including Patrick Hamlyn and Andrew Clarke in Australia, formed an Eternity mailing list [15]. The science of polyform solving increased vastly over the ensuing months. The Eternity mailing list, the breadth-first search algorithm, and many clever computer programs cracked the difficult puzzle. Alex Selby, with help from Oliver Riordan, was the first to find a solution, followed by mere weeks by Guenter Stertenbrink. Mon-ckton's intended solution remains unknown; he planned another contest to find his specific solution via a series of clues, which would be substantially more difficult to solve.

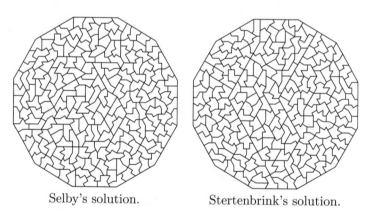

Selby's solution.   Stertenbrink's solution.

Because of lack of sales (mainly in the United States, where the puzzle never caught on) and the short amount of time (less than a year) needed to solve it, Eternity wound up being a financial failure. Monckton lost his mansion and a lot of money, but he made sure the first solver, Alex Selby, got paid. It was a fair contest. Monckton was a thoroughly good sport through it all, and helped out the cause of recreational mathematics in the process.

Powerful solving techniques had been created, so Hamlyn started a new mailing list devoted to polyforms in general [16]. One of the aspects that has been much studied is the creation of aesthetically pleasing shapes from *complete* sets of polyforms.

For polyominoes, the obvious shapes to look for are rectangles. It's well known that the 12 pentominoes can be fit together into a 6×10 rectangle. They also fit into rectangles of size 5×12, 4×15, and 3×20 [17, p. 130]. The 35 hexominoes, on the other hand, cannot form a rectangle. The impossibility of this is based on a checkerboard-coloring parity argument: If you alternately color the squares that compose each hexomino, say Black and White, then 24 of the 35 hexominoes get three of each color, but 11 of them—an *odd* number—get more of one than the other, and this makes it impossible to balance the coloring of the complete set to have equal amounts of each color [17, p. 136].

However, Hamlyn has shown that the hexominoes *and trominoes* together can make not only rectangles, but six 6×6 squares, as shown below.

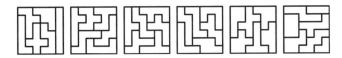

Along similar lines, Hamlyn found that the 108 heptominoes form 12 8×8 squares, each with a hole in the same place. Furthermore, his solution can be three-colored in such a way that three identical, 11×23 rectangles can be made with each color.

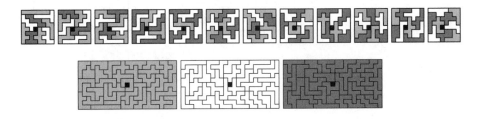

Turning to polyiamonds, we get some striking results:

- The 4 pentiamonds can cover an icosahedron.

- The 12 hexiamonds can make a diamond with side length 6.

- The 24 heptiamonds can make a parallelogram with sides of length 3 and 28.

- The 66 octiamonds can make a parallelogram with sides of length 4 and 66.

- The hexiamonds and octiamonds combined make a hexagon of side length 10. In this packing, the octiamonds form an "octiamond ring" around an "iamond hex" [19].

Michael Dowle sent me another nice pattern with the 66 octiamonds:

After studying octiamonds, and believing my money would be safe (this was before Eternity came out), I made a $100 challenge: Make 11 identical shapes with the full set of octiamonds. Hamlyn solved it:

Hamlyn, who is a windsurfer, also used octiamonds for a sailboard design:

The 24 heptiamonds can make a side-13 triangle with a hole in the center. Likewise, the 66 octiamonds can make a side-23 triangle with a hole in the center. The 160 enneiamonds (polyiamonds with 9 equilateral triangles) cannot form a single triangle with a hole. That's because, unlike $24 \times 7 + 1 = 169$ and $66 \times 8 + 1 = 529$, $160 \times 9 + 1 = 1441$ is not a square. (The number of unit triangles in a side-n triangle is n^2.) However, Andrew Clarke found an arrangement of the 160 enneiamonds into four side-19 triangles, each with a hole at the center:

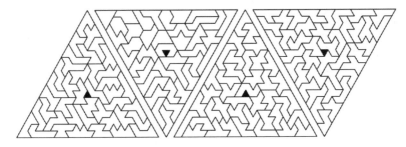

Hamlyn has found an arrangement into five side-17 triangles, each with a hole. (Note: $(1440 + 4)/4 = 19^2$ and $(1440 + 5)/5 = 17^2$. One might similarly ask if it's possible to arrange the enneiamonds into 12 side-11 triangles, each with a hole—or 10 side-12 triangles with no holes.)

Hamlyn has also put the 448 deciamonds in a strip:

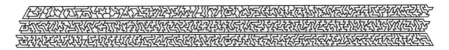

New types of polyforms have been created, inspired in part by Eternity. For example, Brendan Owen suggested polykites, where kites are created by cutting an equilateral triangle into three pieces as shown below.

Hamlyn found that the 84 "unholey" hexakites can be arranged into 7 hexagons:

Another set of polyforms are the halfominoes, formed by connecting n squares and a single diagonally cut half square. For $n = 0, 1, 2$, etc., there are 1, 1, 4, 16, 35, etc. n.5-ominoes. The 35 4.5-ominoes can make five clipped rectangles, like old-style computer punch cards:

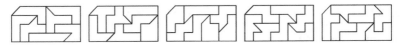

I close with one final polyform composition. A unit square, an isosceles right triangle with legs of length 1, and a right triangle with legs of length 1 and 2 can be fit together along their unit sides in 32 different ways. Each piece has area 2.5, so the entire set has area 80. It turns out these pieces can be arranged into a square:

References

[1] Andrew Clarke's Poly Pages: http://clarkjag.idx.com.au/

[2] Geometry Junkyard: http://www.ics.uci.edu/~eppstein/junkyard/polyomino.html

[3] Roel Huisman: http://www.homepages.hetnet.nl/~eomer/pforms/start.html

[4] Kadon Enterprises: http://gamepuzzles.com/

[5] Rodolfo Kurchan's Puzzle Fun: http://www.eldar.org/~problemi/pfun/pfunmain.html

[6] Alexandre Owen Muiz: http://xprt.net/~munizao/mathrec/pentcol.html

[7] Brendan Owen: http://members.optusnet.com.au/polyforms/

[8] Ed Pegg Jr: http://www.mathpuzzle.com

[9] Michael Reid: http://www.math.ucf.edu/~reid/Polyomino/index.html

[10] Alex Selby: http://www.archduke.demon.co.uk/eternity/index.html

[11] Torsten Sillke: http://www.mathematik.uni-bielefeld.de/~sillke/

[12] Miroslav Vicher's Puzzles Pages: http://alpha.ujep.cz/~vicher/puzzle/

[13] Aad van de Wetering: http://home.wxs.nl/~avdw3b/aad.html

[14] Livio Zucca: http://www.geocities.com/liviozuc/index.html

[15] Eternity group: http://groups.yahoo.com/group/Eternity/

[16] Polyforms group: http://groups.yahoo.com/group/polyforms/

[17] Martin Gardner, Mathematical Puzzles and Diversions. New York: Simon and Schuster, 1959.

[18] N. J. A. Sloane, editor, The On-Line Encyclopedia of Integer Sequences, http://www.research.att.com/~njas/sequences/. In particular, sequence A000105 counts polynominoes, A000577 counts polyiamonds, and A056842 counts polydrafters.

[19] Kadon Enterprises, Inc. Essential Polyforms: http://www.gamepuzzles.com/esspoly.htm

Another Pentomino Problem

Solomon W. Golomb

The Problem

For each of the twelve pentominoes, we ask where it can be placed on a 5×7 board in such a way that the rest of the board can be tiled with ten right trominoes $\left(\begin{array}{c}\rule{0pt}{0pt}\end{array}\right)$. There are at least three locations, distinct with respect to the symmetry group of the 5×7 rectangle, for each pentomino.

Methodology

Consider the 12 "dotted" squares shown. Since a right tromino can cover at most one dot, the ten right trominoes will cover at most ten dots altogether.

Solomon W. Golomb, University Professor and Viterbi Professor of Communications at the University of Southern California, is an elected member of the National Academy of Sciences and the National Academy of Engineering. His numerous honors and awards for his seminal work in communication theory include the Shannon Award of the IEEE Information Theory Society and the IEEE Hamming Medal. He is well-known as the inventor of polyominoes and authored several articles in Martin Gardner's *Mathematical Games* column in Scientific American as well as his book *Polyominoes*. He is also the author of several puzzle columns.

Therefore the pentomino must cover at least two of the dotted squares.

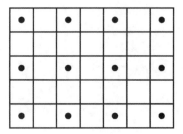

Solutions

The next page gives three solutions for each of the 12 pentominoes. (We consider solutions distinct only if they involve different locations of the pentomino on the 5 × 7 board, and not merely a rearrangement of the right trominoes.) Since the 2 × 3 rectangle $\left(\vcenter{\hbox{⊞⊞}} \right)$ can be tiled with right trominoes in two ways, we leave it undivided in the pictures of the solutions. The reader is encouraged to find the remaining inequivalent solutions for those pentominoes that have more than three successful placements. For example, the P-pentomino $\left(\vcenter{\hbox{⊞}} \right)$ has additional solutions, while the X-pentomino $\left(\vcenter{\hbox{⧂}} \right)$ has only the three that are shown.

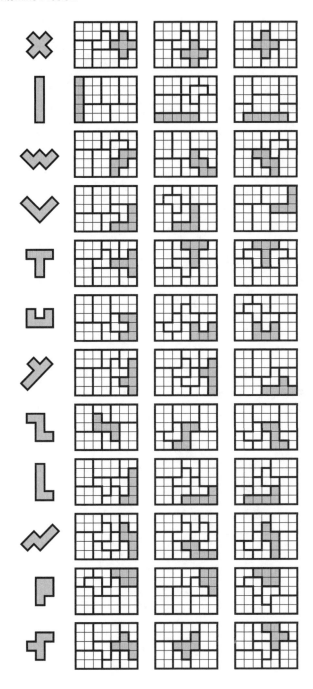

Polyomino Number Theory (III)

Uldis Barbans, Andris Cibulis, Gilbert Lee, Andy Liu,
and Robert Wainwright

Polyominoes are connected plane figures formed by joining unit squares edge to edge. A polyomino A is said to **divide** another polyomino B if a copy of B may be assembled from copies of A. We also say that A is a **divisor** of B, B is **divisible** by A, and B is a **multiple** of A. A polyomino is said to be a **common multiple** of two other polyominoes if it is a multiple of both. If two polyominoes have common multiples, they are said to be **compatible**. A **least common multiple** of two compatible polyominoes is a common multiple with minimum area.

In our previous papers [1, 2], we investigated the compatibility of polyominoes of order 3, 4, and 5, except between two pentominoes. Following the "penta" theme of the Fifth Gathering for Gardner, we present here our findings on compatibility among the pentominoes.

The problem is surprisingly intriguing. Our results are summarized in the chart in Figure 2, but to whet the reader's appetite, we present

Uldis Barbans (undergraduate student) and **Andris Cibulis** are at University of Latvia in Riga. **Gilbert Lee** (graduate student) and **Andy Liu** are at the University of Alberta in Edmonton, Canada. **Robert Wainwright** is a mathematics teacher living in New Rochelle, New York, and edited *Lifeline*, a quarterly newsletter for enthusiasts of John Conway's Game of Life.

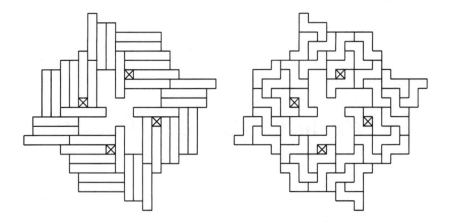

Figure 1. A common multiple of the I-pentomino and the Z-pentomino.

in Figure 1 the smallest common multiple of the I-pentomino and the Z-pentomino that we can find. We have no idea how close it is to being a least common multiple.

In Figure 2 the pentominoes, other than the X-pentomino, are featured along the main diagonal in the chart. If there is a figure in the ith row and the jth column, it shows the construction from the ith pentomino of a least common multiple with minimal area 10 of the ith and jth pentominoes. If there is a blank, it means that there is a common multiple of area exceeding 10, shown in Figure 3. All are least common multiples, which we verify with the aid of computers, except possibly those of the following pairs: (I,F), (I,W), (I,U), (I,Z), and (T,W).

For the X-pentomino, we can only find common multiples with F, Y, P, T, N, and L. The first five are shown in Figure 4, and are in fact least common multiples except possibly for the pair (X,N).

In our previous paper [2], we have proved that the X-pentomino is compatible with neither the I-tromino nor the I-tetromino. The same argument shows that it is not compatible with the I-pentomino either. We now modify our approach used to prove that the X-pentomino is not compatible with the O-tetromino to prove that it is not compatible with the U-pentomino.

Suppose a common multiple M exists. Place it inside the smallest rectangle with edges parallel to the grid lines. A unit square in the common multiple is called a support square if it touches an edge of this minimal rectangle. Two support squares of M from the same copy of the U-pentomino are either adjacent or have exactly one empty square in between. However,

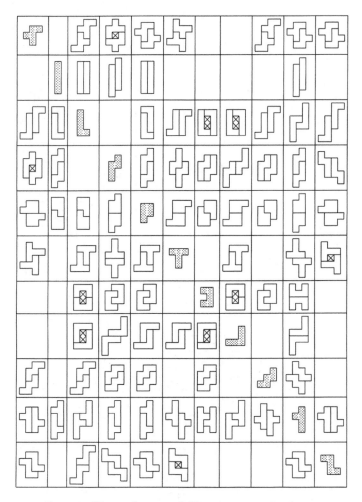

Figure 2. Chart of compatibility among pentominoes.

because M is also a multiple of the X-pentomino, any two of its support squares must have at least two empty squares in between. This is a contradiction.

We claim that the X-pentomino is not compatible with the V-pentomino either. The latter must be placed against the bottom edge of the minimal rectangle as shown in Figure 5, covering the square marked 1. The X-pentomino which covers the square marked 1 must also cover the square marked 2. The latter can only be covered by another V-pentomino as shown. If the square marked 3 is to be covered by the V-pentomino, then

Figure 3. Common multiples of pentominoes of area exceeding 10.

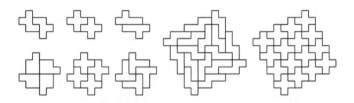

Figure 4. Common multiples of X with F, Y, P, T, and N.

either it or the square marked 4 cannot be covered by X-pentominoes. Hence the squares marked 5 and 6 must be covered by X-pentominoes as shown. We have now recreated the initial situation, with the square marked 4 playing the role of the square marked 1. This justifies our claim.

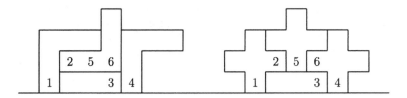

Figure 5. The X-pentomino is not compatible with the V-pentomino.

Figure 6. A common multiple of the X-pentomino and the L-pentomino.

We have not yet determined whether the X-pentomino is compatible with either the W-pentomino or the Z-pentomino, but saving the best till last, we exhibit in Figure 6 the smallest common multiple of the X-pentomino and the L-pentomino that we can find.

References

[1] U. Barbans, A. Cibulis, G. Lee, A. Liu and B. Wainwright, Polyomino Number Theory (II), in "Mathematical Properties of Sequences and other Combinatorial Structures", ed. J. S. No et al., Kluwer Academic Publishers, Dordrecht, 2003, 93–100.

[2] A. Cibulis, A. Liu and B. Wainwright, Polyomino Number Theory (I), *Crux Mathematicorum*, 28 (2002) 147–150.

[3] S. W. Golomb, Normed Division Domains, *American Mathematical Monthly*, 88 (1981) 680–686.

A Golomb Gallimaufry

Jeremiah Farrell, Karen Farrell, and Thomas Rodgers

Solomon Golomb is well known as the inventor of polyominoes [1]. He is also well known as a recreational linguist, and his abilities extend into a variety of word puzzles. He has written several articles for *Word Ways, the Journal of Recreational Linguistics*. One of these, "Amalgamate, Chemist!" [2], is an anagram of Martin Gardner's *Scientific American* column "Mathematical Games." Douglas Hofstadter coined another anagram, "Metamagical Themas," from the same letters when he took over the column from Gardner in 1981. Golomb (who contributed several "Games" columns himself over the years [3]) went on to list 100 anagrams of "Mathematical Games."

In tribute to Sol, we have created a combination of polyomino problems and wordplay, based on the crossword-like diagram below. Diagram 1 is a 9 × 9 grid that has been partially blacked out by four straight trominoes.

In addition to their admiration for Solomon W. Golomb and Martin Gardner, the co-authors share in the excitement of collecting antiquarian dictionaries, puzzles, and mathematical recreations of all types.

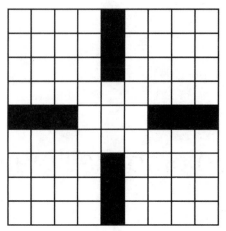

Diagram 1.

Problem 1. Show that Diagram 1 can be completely covered with 23 more straight trominoes (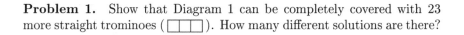). How many different solutions are there?

Problem 2. Show that it is impossible to complete the tiling of Diagram 1 with 23 right trominoes $\left(\rule{0pt}{2ex}\right)$.

Problem 3. Suppose that we have 14 right trominoes and ten straight trominoes in addition to the four straight trominoes already placed. Show that Diagram 1 can be tiled with these.

Problem 4. Suppose that we have 13 P-pentominoes $\left(\rule{0pt}{2ex}\right)$, one tromino (of either type), and one monomino. Is it possible to complete the tiling of Diagram 1?

Problem 5. Suppose that we play a two-player game by alternately placing trominoes on Diagram 1 until someone has no play and loses. Show that, with an unlimited supply of pieces (straight and right), the first player has a winning strategy.

Problem 6. Suppose that the supply of pieces in the game of Problem 5 is limited to 14 right and ten straight trominoes. Does the first player still have a winning strategy?

Problem 7. As a crossword puzzle, Diagram 1 has two 3-letter words, 24 4-letter words, and four 9-letter words. We will give you the clues, but won't tell you which word goes where. Nonetheless, there is a unique solution, up to a reflection about the main diagonal. For the 3- and 9-letter entries, complete (if cryptic) clues are given; the 4-letter entries are all anagrams of the clues. (The clues are all meaningful words. BERH, for example, is a city in Mongolia.)

3's   Reg. test for a Senior
       Golden Roo?

4's   AARE,   AGRA,   BERH,   DOME,   GOAL,   GOBS,
       GORE,   HAUT,   HEST,   MAID,   MAIL,   MEAT,
       NAGS,   NOIR,   NOME,   NOSE,   OAST,   RENO,
       RETE,   SPAD,   SUBS,   TAOS,   TOOT,   TROT

9's   A log blooms again, man!
       Hag's doggy a tall tale with a tail?
       Steal meat for an old spouse?
       I? Monopoly? No, too square.

Solutions

Problem 1. The middlemost square must be covered by one piece, say a horizontal. The squares above and below it must also be covered by horizontals, and each must be the middle square of the horizontal that covers them; otherwise there would be a grid of 14 squares on one side and 16 on the other, neither of which can be covered by trominoes, since neither 14 nor 16 is divisible by 3. Thus the middle 3×3 portion of Diagram 1 can be covered in only two ways, leaving four identical grids in the corners. It is easy to see that each of these can be covered in just 4 different ways. Thus the total number of solutions is $2 \cdot 4 \cdot 4 \cdot 4 \cdot 4 = 512$.

Problem 2. Each 4×4 corner takes five right trominoes with one blank, which must be the interior corner. Hence we must cover the 3×3 center with right trominoes and that is clearly impossible.

Problem 3. There are many ways of doing this. (We do not know how many.) It may be that the 3×3 center is necessarily filled with three straight trominoes. After this start, it is easy to tile the rest.

Problem 4. Tile each of the four corners using three P-pentominoes, leaving the inner corner blank. This leaves the 3 × 4 interior, which can be covered with one P-pentomino and one monomino and either of the two trominos.

Problem 5. Cover up the middle square with a straight piece, then play a *tweedle-dee/tweedle-dum* strategy: whatever your opponent plays, play the same but rotated 180° around the center.

Problem 6. To employ a tweedle-dee/tweedle-dum strategy, we would have to use a straight tromino to cover the center square. But this leaves nine straights and the first player would not always have a straight to mirror the second player's play. So it is an open problem to determine who should win this game.

Problem 7.

B	O	G	S		S	E	T	H
U	T	A	H		T	R	E	E
S	T	O	A		A	G	A	R
S	O	L	G	O	L	O	M	B
		G	R	E				
P	O	L	Y	O	M	I	N	O
A	M	I	D		A	R	E	A
D	E	M	O		T	O	R	T
S	N	A	G		E	N	O	S

References

[1] Golomb, Solomon, *Polyominoes.* Revised and Expanded Edition. Princeton, NJ: Princeton University Press, 1994.

[2] Golomb, Solomon, "Amalgamate, Chemist," *Word Ways, the Journal of Recreational Linguistics*, Vol. 17, No. 1, February, 1984, pages 20-23.

[3] Golomb, Solomon, "Mathematical Games," *Scientific American*, May and December 1957, November and December 1960, May and December 1963,

October and November 1965, October 1967, March and April 1972, and August and September 1975.

Part V

Braintaunters

The Panex Puzzle

Mark Manasse, Danny Sleator, Victor K. Wei, Nick Baxter

Introduction

The Panex puzzle is a one-person board game created by Toshio Akanuma and manufactured by TRICKS Co., Ltd. of Tokyo, Japan. On first sight, the puzzle reminds one of the Tower of Hanoi. A little thought reveals that they are intriguingly different.

The puzzle consists of a flat board with three vertical tracks laid in the board with a horizontal track at the top connecting the three. The board is made of two flat pieces fastened together by screws, with intricate shapes hidden between them. Rectangular tile pieces can be moved along and inside the tracks, but cannot be lifted out of the board, nor rotated. Thus, for example, one tile cannot fly over another tile. For the Silver version (see Figure 1), there are ten tiles with blue markings, and ten with

Mark Manasse is a senior researcher at Microsoft. **Danny Sleator** is professor of computer science at Carnegie Mellon University. **Victor K. Wei** is professor of information engineering at the Chinese University of Hong Kong. **Nick Baxter** is a mathematician and puzzle expert with a specialty in sliding block puzzles.

This chapter is based on a paper originally written in 1983 by M. Manasse, D. Sleator, and V. K. Wei [6], but never formally published. The original paper, a description of the authors' search program, and other related information is at N. Baxter's Panex Resources [2].

Figure 1. Panex Silver.   Figure 2. Sample position.   Figure 3. Panex Gold.

orange markings. The markings of the same color are such that they form a narrow, ten-story high triangle-shaped tower, with each tile taking up one floor (or layer). For the Gold version (see Figure 3), each tower is made up of tiles that appear identical, but otherwise is constructed identically to the Silver version.

The rule of the game is such that the tile with the ith highest floor can only stay in layers 0 through i in any track. (The top layer of a track, where it is connected to another track, is numbered 0. The *cul-de-sac* of a track is numbered 10.) This rule is enforced by the mechanics of the Panex board (see Figure 2). There is a tongue (or tang) underneath each tile which decreases in size from first to the tenth tiles. The throat of the track narrows from layer 1 to layer 10. In summary, the rules of the game are

1. The tiles can only move along the tracks. They cannot fly over one another.

2. The tile with the ith highest floor of the triangular tower can only stay in track layers 0 through i.

Initially, the blue tiles are stacked on the left track, and the orange tiles are on the right track. There are two goals to be played on the Panex puzzle. The first is to transfer one tower from a side track to the center track. The second is to exchange two side towers.

Although the TRICKS Company implemented the Panex puzzle with towers of height ten only, we can treat the general puzzle with towers of height n. Let $T(n)$ denote the minimum number of moves to transport a tower of height n from a side track to the center track, and let $X(n)$ denote the minimum number of moves to exchange two towers on the side tracks. In this paper, we will give a formula for $T(n)$ and give upper and lower bounds on $X(n)$. Note that the displacement of a tile is counted as one move. It does not matter how far along the tracks the tile travels, or how many turns it makes.

For $n \geq 3$, the minimum number of moves to transport a tower is

$$T(n) = c_1(1 + \sqrt{2})^n + c_2(1 - \sqrt{2})^n + \frac{(-1)^n}{2} - 1 \quad ,$$

where

$$c_1 = \frac{7}{4}(-1 + \sqrt{2}), \quad \text{and} \quad c_2 = \frac{7}{4}(-1 - \sqrt{2}).$$

As for the minimum number of moves to exchange towers on the side tracks, $X(n)$, we have obtained the exact value of $X(n)$ for $1 \leq n \leq 8$ by exhaustive computer search. For $n \geq 5$, we have the following upper and lower bounds, $L(n) \leq X(n) \leq U(n)$. The upper bound is given by

$$U(n) = c_3(1 + \sqrt{2})^n - 8n - 2 + c_4(1 - \sqrt{2})^n,$$

where

$$c_3 = \frac{7}{2}(7 - 4\sqrt{2}), \quad \text{and} \quad c_4 = \frac{7}{2}(7 + 4\sqrt{2}).$$

The lower bound is given by

$$L(n) = 4(T(n) + T(n - 1) - 2).$$

A tabulation of the results for $n \leq 10$ is given in Table 1.

We first present the optimum algorithm for transferring a single tower, then an algorithm for exchanging two towers in $U(n)$ moves.

The Optimum Sequence of Moves for Transferring Towers

In this section, we will give an algorithm for transferring a tower from a side track to the center track and prove that the moves used are the fewest possible. Specifically, we shall focus on the task of transferring the blue

n	$T(n)$	$X(n)^1$	$U(n)$	$L(n)$
1	1	3		
2	3	13		
3	9	42		
4	24	128		
5	58	343	343	320
6	143	881	881	796
7	345	$2,189^2$	2,189	1,944
8	836	$5,359^2$	5,359	4,716
9	2,018	?	13,023	11,408
10	4,875	?	31,537	27,564

Table 1.

tower from left to center. We will accomplish the transfer problem by a series of *atomic operations*. Each atomic operation is a fixed sequence of moves that achieves a sub-goal toward the eventual goal of tower transfer. Initially, we will prohibit the movement of orange tiles. Later, we will show that the removal of this restriction does not help.

Let T_n denote the operation of transferring a tower of height n from a side track to the center track. Let S_n denote the operation which takes the position where an $(n-1)$-tower is in the center and the nth tile is alone on the side, to the position where the $(n-1)$-tower is on the side and the nth tile is alone in the center. For example, the operation S_6 is illustrated below.

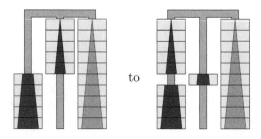

to

The task T_n can be accomplished in $T(n)$ but not fewer moves. The task S_n can be accomplished in $S(n)$ but not fewer moves. We use the

[1]These $X(n)$ values are results verified by the authors' computer search program; see [2].

[2]Verified first by David Bagley in 2002

letter S to name the second operation because it *sinks* the #n tile to the bottom of the center track.

There are two operations similar to S_n that we will assume take the same minimum number of moves. This assumption will be justified later. In the meantime, these two operations will also be referred to as S_n. These situations are caused by particular combinations on the lower portion of the board. Ambiguity should not arise from reading the context.

For the first variation of S_n, there is an single extra tile at height n in the center track. (Illustrated for S_6.)

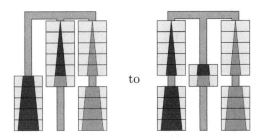

For the second variation of S_n, tile #$(n+i)$, where $i \geq 1$, instead of tile #n, is sunk to the bottom of a tower of height $n-1$ in the center track. (Illustrated for S_6 and $i = 1$.)

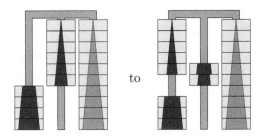

Let $\bar{T}_n$ denote the operation of transporting a tower of height n from the side track to the center, but with the first tile ending up in the opposite corner position. (Illustrated for $\bar{T}_6$.)

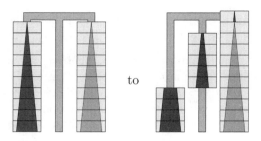

Assume that $\bar{T}_n$ can be accomplished in $\bar{T}(n)$ but not fewer moves. Let $\bar{S}_n$ denote the operation of S_n, but with the 1st tile starting in the opposite corner. (Illustrated for $\bar{S}_6$.)

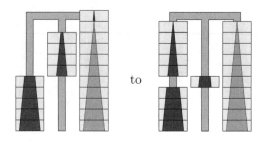

Also assume $\bar{S}_n$ can be accomplished in $\bar{S}(n)$ but not fewer moves. Similar assumptions are made for $\bar{S}_n$ as for S_n.

Let T_n^{-1}, S_n^{-1}, $\bar{T}_n^{-1}$, and $\bar{S}_n^{-1}$ denote the inverse operations of T_n, S_n, $\bar{T}_n$, and $\bar{S}_n$, respectively. Clearly, T_n^{-1} (and S_n^{-1}, $\bar{T}_n^{-1}$, and $\bar{S}_n^{-1}$ respectively) uses the same number of moves as T_n (and S_n, $\bar{T}_n$, and $\bar{S}_n$ respectively).

Without much effort, the reader can verify that

$$T(1) = \bar{T}(1) = 1, \; T(2) = S(2) = 3, \; \bar{T}(2) = \bar{S}(2) = 2, \; T(3) = \bar{T}(3) = 9,$$
$$\text{and } S(3) = \bar{S}(3) = 8.$$

For $n \geq 4$, assume we know the shortest sequence of moves to accomplish $T(i)$, $\bar{T}(i)$, $S(i)$, $\bar{S}(i)$ for all $i \leq n$, then the task T_n can be accomplished by applying T_{n-1}, S_{n-1}, a special sequence of four moves, and T_{n-2}, in that order. The positions after each stage are shown, for $n = 6$.

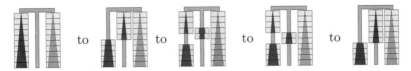

Another way to accomplish T_n is by applying $\bar{T}_{n-1}$, $\bar{S}_{n-1}$, a special sequence of four moves, and T_{n-2}, in that order. The intermediate positions reached at each stage is shown below, for $n = 6$.

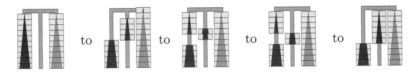

Therefore, the minimum number of moves to achieve T_n is bounded by

$$T(n) \leq \min\left\{T(n-1) + S(n-1); \bar{T}(n-1) + \bar{S}(n-1)\right\} + 4 + T(n-2).$$

Similarly, the task $\bar{T}_n$ can be accomplished by combining smaller tasks, and we have

$$\bar{T}(n) \leq \min\left\{T(n-1) + S(n-1); \bar{T}(n-1) + \bar{S}(n-1)\right\} + 4 + \bar{T}(n-2).$$

The task S_n can be accomplished by applying S_{n-1}, a special sequence of four moves, S_{n-1}^{-1}, and T_{n-2}^{-1}, in that order. The positions reached after each stage are shown below, for $n = 6$.

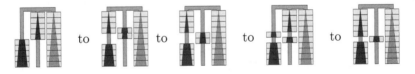

Another way to accomplish S_n is to combine S_{n-1}, a special sequence of four moves, $\bar{S}_{n-1}^{-1}$, and $\bar{T}_{n-2}^{-1}$, in that order. Therefore the minimum number of moves to accomplish S_n is bounded by

$$S(n) \leq S(n-1) + 4 + \min\left\{S(n-1) + T(n-2); \bar{S}(n-1) + \bar{T}(n-2)\right\}.$$

Similarly, the task $\bar{S}_n$ can be accomplished by combining smaller operations, and we have

$$\bar{S}(n) \leq \bar{S}(n-1) + 4 + \min\left\{S(n-1) + T(n-2); \bar{S}(n-1) + \bar{T}(n-2)\right\}.$$

Next, we will show that the equalities hold in all four cases.

In any shortest sequence of moves accomplishing T_n, $n \geq 4$, consider the position just before the $\#n$ tile is moved for the first time. In order for the $\#n$ tile to make a meaningful move, we must have either of the following two positions, illustrated here for $n = 6$.

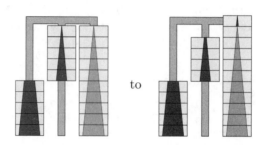

This is because the $\#n$ tile must move all the way to the top and around the corner to be meaningful, and these are the only two possible placements of the other tiles to make the move feasible. These two positions are called an *unavoidable* set of positions. The operation from the initial position to the unavoidable position on the left is exactly T_{n-1}, and the operation from the initial position to the unavoidable position on the right is exactly $\bar{T}_{n-1}$. Therefore, the first part of any shortest sequence of moves to accomplish T_n must be a shortest sequence for either T_{n-1} or $\bar{T}_{n-1}$.

Next, consider the position just after the $\#(n-1)$ tile is moved for the last time in a shortest sequence. There are two possible positions, as shown below for $n = 6$.

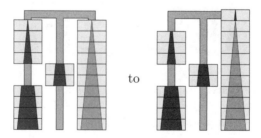

But there is no meaningful sequence of moves that can lead into the position on the right, considering the $\#(n-1)$ tile has just been moved. Therefore, the position on the left is an unavoidable position for any shortest sequence. The task from this position to the end is exactly T_{n-2}. Therefore, the last part of any shortest sequence of moves for accomplishing T_n must be a shortest sequence for accomplishing T_{n-2}.

Tracing backwards from this unavoidable position, we see that four moves earlier, we must have been in the position

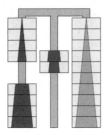

in any shortest sequence. This position is also unavoidable. The sequence of moves before reaching this position must be a shortest sequence for S_{n-1} or $\bar{S}_{n-1}$. Summarizing the arguments, we have

$$T(n) = \min\{T(n-1) + S(n-1); \bar{T}(n-1) + \bar{S}(n-1)\} + 4 + T(n-2)$$

for $n \geq 4$. Similarly, we can argue on unavoidable positions and prove that, for $n \geq 4$,

$$\bar{T}(n) = \min\{T(n-1) + S(n-1); \bar{T}(n-1) + \bar{S}(n-1)\} + 4 + \bar{T}(n-2).$$

In any shortest sequence of moves that accomplishes S_n, $n \geq 4$ consider the last time the $\#(n-1)$ tile is moved. We must be in either of the following unavoidable set of positions, illustrated for $n = 6$.

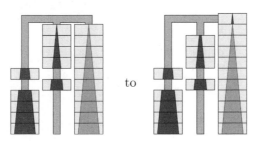

to

Focusing on the time when tile $\#n$ and tile $\#(n-1)$ change relative positions, we find the following two unavoidable positions, illustrated for $n = 6$. We must reach the unavoidable position on the left, and four moves later, reach the unavoidable position on the right.

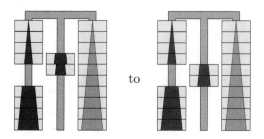

Therefore, we have

$$S(n) = S(n-1) + 4 + \min\{S(n-1) + T(n-2); \bar{S}(n-1) + \bar{T}(n-2)\}$$

for $n \geq 4$. Similarly, we have

$$\bar{S}(n) = \bar{S}(n-1) + 4 + \min\{S(n-1) + T(n-2); \bar{S}(n-1) + \bar{T}(n-2)\}$$

for $n \geq 4$.

Based on the values for small n, and the four equations, we can derive that, for $n \geq 4$,

$$\bar{T}(n) = \begin{cases} T(n), & n \text{ odd} \\ T(n) - 1, & n \text{ even} \end{cases}$$

and

$$\bar{S}(n) = S(n) = \bar{T}(n) - 1.$$

Hence,

$$\begin{aligned} \bar{T}(n) &= \bar{T}(n-1) + \bar{S}(n-1) + 4 + \bar{T}(n-2) \\ &= 2\bar{T}(n-1) + \bar{T}(n-2) + 3. \end{aligned}$$

Solving by the techniques of difference equations, we obtain

$$\bar{T}(n) = c_1(1 + \sqrt{2})^n + c_2(1 - \sqrt{2})^n - \frac{3}{2}$$

for $n \geq 4$, where

$$c_1 = \frac{7}{4}(-1 + \sqrt{2}), \quad \text{and} \quad c_2 = \frac{7}{4}(-1 - \sqrt{2}).$$

The equation for $T(n)$ can be obtained easily, and has been given in the introduction. The values of $T(n)$, $\bar{T}(n)$, $S(n)$, and $\bar{S}(n)$ are shown in Table 2.

n	$T(n)$	$\bar{T}(n)$	$S(n)$	$\bar{S}(n)$
1	1	1		
2	3	2	3	2
3	9	9	8	8
4	24	23	22	22
5	58	58	57	57
6	143	142	141	141
7	345	345	344	344
8	836	835	834	834
9	2,018	2,018	2,017	2,017
10	4,875	4,874	4,873	4,873

Table 2.

Earlier, we have assumed that two variations of the operation S_n take the same number of moves as S_n. Judging from the algorithm that we presented for accomplishing S_n, and the arguments based on unavoidable positions, we see that these assumptions are indeed true. Earlier, we have also prohibited the movement of the orange tiles. Again, judging from the algorithm and the unavoidable positions, we see that the removal of this restriction does not help shorten the sequence of moves needed to accomplish T_n.

A Good Algorithm for Exchanging Two Towers

In this section we present an algorithm for exchanging two towers on the side tracks.

We need some more detailed notations to present our algorithm. For $Y = T, \bar{T}, S,$ or $\bar{S}$, let $_LY_n$ denote the operation of performing the task on the left and center tracks, and let $_RY_n$ denote the operation of performing the task on the right and center tracks. The combination of colors of the tiles on the tracks is irrelevant. For example, the following operation is denoted $_RT_4$.

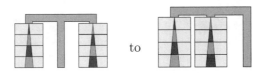

As before, Y^{-1} denote the inverse operation, and the sink task S_n requires the same number of moves even if the sinking tile is $\#(n+i)$, $i > 0$, and/or the ith layer of the center track is already occupied. For example, the following operation is also denoted $_LS_5$.

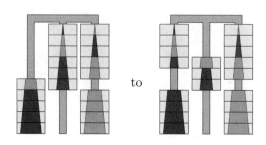

to

The reader can easily verify that $X(1) = 3$ and $X(2) = 13$. The following sequence of operations accomplishes X_3 in 42 moves:

$$_L\bar{T}_2^+ {}_L\bar{S}_3 + {}_R\bar{T}_2 + Y + Z$$

where Y is the following operation which requires 15 moves:

to

and Z is the following operation which also requires 15 moves:

to

In Y, intermediate positions after 3, 6, 9, and 12 moves are

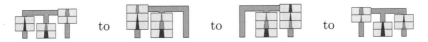

In Z, intermediate positions after 3, 6, 9, and 12 moves are

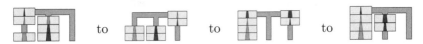

The reader can easily figure out the complete details of Y and Z.

The following sequence of operations accomplishes X_4 in 128 moves:

$$_L\bar{T}_3^+{}_L\bar{S}_3 + 4 + {}_R\bar{T}_2 + \bar{Y} + {}_L\bar{T}_2^{-1} + {}_R\bar{T}_2 + {}_R\bar{S}_4$$
$$+ {}_L\bar{S}_4^{-1} + W + {}_RS_3 + 4 + {}_R\bar{S}_3^{-1} + {}_R\bar{T}_3^{-1}$$

where $\bar{Y}$ is the following operation which requires 15 moves:

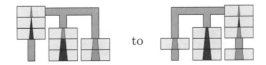

and W is the following operation which requires 13 moves:

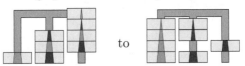

The operation "4" swaps two pieces in the center track in four moves. The first 12 moves of $\bar{Y}$ and Y are identical. The reader can easily figure out the last three moves of $\bar{Y}$. In W, the positions reached after 3, 6, 9, and 11 moves are

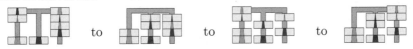

For convenience, define C_n as the following combination of moves:

$$_L\bar{T}_{n-1}^{-1} + {}_R\bar{T}_{n-1}\,{}_R\bar{S}_n + {}_L\bar{S}_n^{-1}$$

which moves the $\#i$ tile from the right track to the left (shown for C_6).

For $n \geq 5$, the task X_n is accomplished by the following four series of operations:

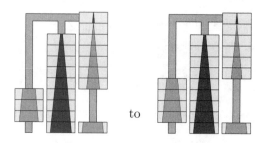

to

1. $_L\bar{T}^+_{n-1}{}_L\bar{S}_{n-1} + 4 + {}_R\bar{T}_{n-2} + {}_R\bar{S}_{n-1} + {}_L\bar{S}^{-1}_{n-1},$

2. $C_{n-2} + C_{n-3} + \ldots + C_5 + C_4,$

3. $_L\bar{T}^{-1}_2 + {}_R\bar{T}_2 + \bar{Y} + {}_L\bar{T}^{-1}_2 + {}_R\bar{T}_2 + {}_R\bar{S}_n,$

4. $_L\bar{S}^{-1}_n + V + {}_R\bar{T}^{-1}_{n-1}.$

The first sequence takes the initial position into the following position, illustrated for $n = 10$.

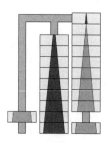

The second sequence moves the orange $\#(n-2)$ through #4 tiles from the right track to the left, resulting in the following position.

The third sequence continues the second sequence with a set of moves roughly equivalent to $C_3 + C_2 + C_1$, except that the #2 tiles are swapped. In doing so, this sequence saves four moves, at the expense of only one additional move later on to correct it. Then $_R\bar{S}_n$ reaches the (roughly) *half-way position,*

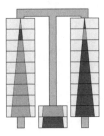

in which all tile pairs except the #2 tiles have been exchanged and the #10 tiles sit in the center track.

The final sequence takes the half-way position to the end position. The operation V uses $\bar{T}(n-1) + \bar{S}(n) + 1$ moves to achieve the tasks $_L\bar{T}_{n-1}^{-1}$, $_R\bar{S}_n^{-1}$, and the exchange of the #2 tiles; it is $_L\bar{T}_{n-1}^{-1}$ followed by $_R\bar{S}_n^{-1}$, with slight modifications to the last part of $_L\bar{T}_{n-1}^{-1}$ and to the first part of $_R\bar{S}_n^{-1}$. According to the algorithm presented in the previous section, the last nine moves of $_L\bar{T}_{n-1}^{-1}$ constitute $_L\bar{T}_3^{-1}$, and the first nine moves of $_R\bar{S}_n^{-1}$ constitute $_R\bar{T}_3$ (for $n \geq 5$). Together, these 18 moves accomplish the following task.

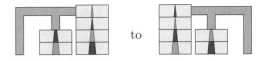

In V, these 18 moves are replaced by 19 moves which take

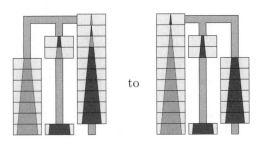

The first 13 moves of this sequence are identical to the operation W described earlier, the reader can easily figure out the remaining 6 moves.

Since $\bar{S}(n) = \bar{T}(n) - 1$, we can accomplish X_n, $n \geq 5$, in $U(n)$ moves, where

$$U(n) = 3\bar{T}(n) + 6\bar{T}(n - 1) + 3\bar{T}(n - 2) + 4 \sum_{i=2}^{n-3} \bar{T}(i) - 2n + 6.$$

Algebraic manipulations produce the formula for $U(n)$ presented in the Introduction. Exhaustive search shows that this upper bound is in fact required for $n \leq 8$. Additionally, we suspect that $X(n) = U(n)$ for all $n \geq 9$.

Arguing on unavoidable positions, we can show that the following sequence is one of the shortest for exchanging only the #n tiles, for $n \geq 5$,

$$_L\bar{T}_{n-1} + {_L}\bar{S}_n + {_R}\bar{T}_{n-1} + {_R}\bar{S}_n + {_L}\bar{S}_n^{-1} + {_L}\bar{T}_{n-1}^{-1} + {_R}\bar{S}_n^{-1} + {_R}\bar{T}_{n-1};$$

we obtain the lower bound

$$\begin{aligned} L(n) &= 4(\bar{T}(n) + \bar{T}(n - 1) - 1) \\ &= 4(T(n) + T(n - 1) - 2). \end{aligned}$$

If desired, the lower bound can be tightened by considering the fewest moves needed to exchange more than one bottom piece.

Acknowledgments

The authors wish to thank Ron Graham for bringing this puzzle to their attention, and F. Chung, Joe Buhler, and Dick Hess for fruitful interactions.

References

[1] D. Bagley, *Panex Applet*, http://gwyn.tux.org/~bagleyd/java/PanexApp.html.

[2] N. Baxter, *Panex Resources*, http://www.baxterweb.com/puzzles/panex.

[3] V. Dubrovsky, Nesting Puzzle, Part 1: Moving Oriental Towers. *Quantum*, Jan/Feb 1996, pp 53-57, 49-51.

[4] E. Henderson, *Panex Puzzle* (level 4 only), http://www.cheesygames .com/panex/ (and uncredited at numerous other sites).

[5] L.E. Hordern, *Sliding Piece Puzzles*. Oxford: Oxford University Press, 1986, pp 144-145, 220.

[6] M. Manasse, D. Sleator, and V. K. Wei, *Some Results on the Panex Puzzle*. Unpublished, 1983.

[7] J. Slocum and J. Botermans, *Puzzles Old & New*. New York: Plenary Publications International, p 135, 1986.

Upstart Puzzles

Dennis Shasha

Of Puzzles Gone Astray

The writer of puzzles often invents puzzles to illustrate a principle. The puzzles, however, sometimes have other ideas. They speak up and say that they would be so much prettier as slight variants of their original selves. What nerve!

The dilemma is that the puzzle-inventor sometimes can't solve those variants. Sometimes he finds out that his colleagues can't solve them either, because there is no existing theory for solving them. At that point, these sassy variants deserve to be called upstarts.

We discuss a few upstarts inspired originally from the Falklands/Malvinas Wars, zero-knowledge proofs, and magic squares. They have given a good deal of trouble to a certain mathematical detective whom I know well.

Puzzles as Antidote to Confusion

When I left college in the late 1970s, I went to work at IBM. Because mine was mostly a theoretical education (Yale engineering was a liberal experi-

Dennis Shasha is a professor of computer science at the Courant Institute of NYU where he does research on pattern discovery in biology and time series. In addition, he has written three puzzle books about the adventures of the famous mathematical detective Dr. Ecco. Shasha also writes the Puzzling Adventures column for Scientific American.

ence), I was quite lost at first when it came to circuit design. Somewhere in college however, I had heard a tactic suggested by Eugene Wigner and William Shockley to the effect that when confronted with a difficult problem, one should look for the simplest non-trivial variation of the problem in order to understand the basic issues. Between that and my love of puzzles, nurtured by my careful reading of Martin Gardner's columns through high schools, I started writing puzzles about digital circuits for myself. One puzzle led to a circuit that has seen wide use in IBM mainframes—a circuit to check that a decoder was working properly.[1] I continued this puzzle-writing habit when I took doctoral courses at Harvard, taking particular inspiration from Michael Rabin's algorithms courses.

The summer after graduate school, my wife and I went trekking in India and I began working on a puzzle book in earnest. Once it was done, though, there was the little matter of getting it published. After many rejections, I sent the manuscript to Martin Gardner quite out of the blue. A few weeks later, Jerry Lyons of W. H. Freeman called me to offer to publish the book, saying that Martin had liked the manuscript.

Dr. Ecco went through two incarnations with Freeman: *The Puzzling Adventures of Dr. Ecco* [6] and the moodier (and regrettably prescient) *Codes, Puzzles, and Conspiracy* [7]. A few years later, the editors of Dr. Dobb's Journal, puzzle freaks by nature, invited me to write the omni-heurist puzzle column aimed at mathematically talented hackers. Variants on those columns appear in *Dr. Ecco's Cyberpuzzles* [8]. Three years later, Scientific American asked me to write the column that has come to be called *Puzzling Adventures*, specifying only that they had to be 500-word puzzles not requiring computer assistance—reading has become a lost art since the three-color, thick magazine days of Martin Gardner. When I write these columns, I am always conscious of Martin's presence. We differ in both style and content, but I want every puzzle to answer at least two of three questions in the affirmative: Can the reader hold the facts in his head? Will the reader learn something useful? Might there be a research generalization?

This article presents three such puzzles that have since turned into upstarts.

Voronoi Game

History shows that most wars are fought between neighbors, often because of different interpretations of borders. With the advent of offshore oil and

---

[1] "Circuits Checking Circuits" in *The Puzzling Adventures of Dr. Ecco* [6].

overfishing, islands often play a big role in disputes because they can define which part of the sea each country owns. In *Codes, Puzzles, and Conspiracy*, I introduced this puzzle as a game played on a line segment between Britain and Argentina called the Territory Game. (The two countries had just fought a war over islands called the Falklands/Malvinas owned by the British but claimed by the Argentines. Oil had been reported in the waters between them.) While writing the book in the summer of 1991, I suggested the two-dimensional version of the puzzle to Philippe Flajolet's group at INRIA in France (where Mordecai Golin was then working) and to Ricky Pollack at NYU. From there, many people have heard it, but it's still unsolved.

Let's start with a simple version for which there is a nice solution. Opponents Red and Blue are given a line segment of unit length. Each has k point-sized stones of its color. Blue lays down all its stones in one round. Then Red lays down its stones but cannot lay them down on top of an existing stone. At the end, any point on the segment that is nearer to a Blue stone than to any Red stone belongs to Blue and vice versa. Does Blue have a winning strategy? A solution is at the end of the article.

This one-dimensional one-round-each version is very specialized. Generalizing it to the case where the players take turns is much harder but has been solved for one dimension [1]. The second generalization is to the plane (or even higher dimensions).

Given a set of point-sized stones in a plane (or subplane), a Voronoi diagram is a tessellation (geometric partitioning) of the plane into polygons such that (i) every stone is in the interior of one polygon and (ii) for every point p in the polygon P containing stone x, p is closer to x than it is to any other stone [5]. Distance is normally based on Euclidean distance.

The Voronoi game is a two-person game (more could play, but the puzzle concerns the two-person variant) on a square that works as follows: you and I each start with k stones, for some k that we agree upon in advance. Suppose the first player gets o-labeled stones and the second gets x-labeled stones. The first player places one stone, then the second player places a stone, and play alternates with each player placing one stone per turn. The player wins whose stones' polygons contain the most territory after all stones have been placed. Figure 1 shows the Voronoi diagram after each player has placed one stone and Figure 2 after the first player has placed his first two stones.

Whereas the Voronoi game is unsolved, there are nice approaches to the one-round version (in which the first player makes all moves before the second player) [2, 3].

A perverse variation of this game makes it even more difficult: Players might play a k/j place-snatch variant in which k stones are laid down as

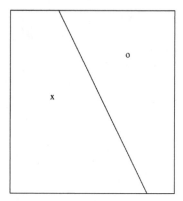

Figure 1.   A two-stone Voronoi diagram. The x owns the left region and the o controls the right region.

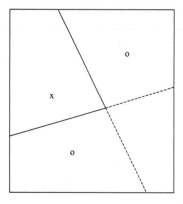

Figure 2.   The game after the first player places two stones. The only segments shown are those between x and o. The dashed segments are fully in the o region.

stated above and then each player removes j stones $(j < k)$ in alternating fashion. Whoever has the most territory with their $k - j$ stones wins.

Prime Geometry

This second puzzle seems to be a problem of computation, but my intuition tells me otherwise. The subject matter is one of the oldest ones in mathematics: prime numbers. Talk about primes and mathematicians grow teary-eyed. Prime numbers are for some God-given, for others the stuff about which we should communicate with aliens. Children grasp the idea quickly, but simply stated hypotheses such as the Goldbach-Euler conjecture (every even number greater than 2 is the sum of two primes) remain unproved. Still, lots is known about the density of primes and other such properties. This puzzle concerns the density of the geometry of primes.

A prime square is a square grid whose rows and columns are prime numbers, but where no two rows are the same nor are two columns the same. An ambidextrous prime square is one which also contains primes when the rows are read from right to left. An omnidextrous prime square is an ambidextrous prime square whose columns are primes when read bottom to top and whose diagonals are also primes in every direction.

Consider the grid of numbers in Table 1.

7	6	9
9	5	3
7	9	7

Table 1. Ambidextrous Prime 3-Square.

Because 769, 953, and 797 are all primes as are the columns 797, 659, and 937, this is a prime square. Because the row reversals 967, 359, and 797 are also primes, this is an ambidextrous prime square. One reason it is not omnidextrous is that 956 is even. Another is that 759 is divisible by 3.

Can you find a prime 4-square that uses 9 distinct digits? A solution is at the end of the article.

The possible questions are many and they are all upstarts as far as I know. There are omnidextrous prime 3-squares that use only 3 distinct digits. What about for omnidextrous prime n-squares? For that matter, for which n are there prime n-squares? ambidextrous prime n-squares? omnidextrous prime n-squares?

Arguments from the density of primes suggest that prime n-squares should become easier to find as n gets larger, but even if that conjecture were true, what does this say about the omnidextrous variants? The reverse may hold.

The Amazing Sand Counter

I wrote this puzzle because I wanted to find the simplest possible zero-knowledge proof. The idea of such proofs is for the Prover to convince the Verifier that the Prover knows something, but the Prover doesn't want to reveal that something [4]. Playing with my three year old at the beach inspired the following puzzle as it first appeared in *Codes, Puzzles, and Conspiracy*:

> A man stepped forward, dressed like a Cossack with high boots and a gilded hat shaped like a mushroom. He bowed with a flourish.
>
> "I am the Amazing Sand Counter. If you put sand into this bucket, I know at a glance how many grains there are," he said (Figure 3). "But I won't tell you."

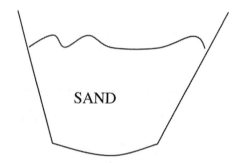

Figure 3.   The Amazing Sand Counter claims to know how many grains are in the bucket just by looking. He wants to prove to you that he knows this without telling you how many there are.

> "Why should we believe you, then?" I said.
>
> At this point, Baskerhound stood up. "That is precisely the question. Can you devise a test by which the sand-counter can prove his skill without telling you anything that you don't already know?"
>
> "Wait. I don't understand," I said. "You want an experiment that follows this form: I put a large amount of sand into the bucket, more than I could possibly count. So, I don't know how many grains there are and neither does the Sand Counter—unless he really has this amazing power. Then I conduct an experiment, which may entail asking him to leave the room or turn away. If he answers some number of questions correctly, I am convinced he knows how many grains of sand are in the bucket, even though he hasn't told me."
>
> "That's right," said Baskerhound. "Any reasonable person should be able to do this test and be convinced. Also, you can

be sure that the Sand Counter has no other amazing power—no X-ray vision or mind-reading capabilities."

The problem is to design such an experiment. Dr. Ecco's solution is at the end of the article.

The upstart variant of this puzzle is the following: suppose the Sand Counter asserted that there were some number n grains in the bucket. Can the Verifier determine whether the Sand Counter consistently tells the truth without doing work that is essentially equivalent to counting the number of grains in the bucket?

The Verifier is allowed to make three kinds of moves: count a grain, divide a pile of grains into approximately equal parts, or ask a question of the Amazing Sand Counter. You would like to limit the number of moves to $\log_2 n$ where n is the number of grains in the bucket. Remember that the Sand Counter's powers are not in question here, merely the truth of his claim and any answer he gives to your subsequent questions. (You win if you figure out the number of grains or catch him lying even once.) As far as I know, this problem is open unless the solver makes additional assumptions [9].

In Conclusion

Puzzles have a personality. That's why some are called nasty, others sweet, and still others fiendish. I'm afraid that I've burdened you with some fiendish ones. Thank Martin.

Solutions

Voronoi Game. Blue would be guaranteed to win if he or she placed the stones at $1/(k+1)$, $2/(k+1)$, ..., $k/(k+1)$.

Prime Geometry. Table 2 shows a prime 4-square that uses 9 distinct digits.

The Amazing Sand Counter. Dr. Ecco solved the puzzle as follows: the Verifier lets the Amazing Sand Counter see the bucket with the sand. Then he asks the Sand Counter to leave the room. The Verifier then takes a small handful of sand from the bucket, being careful not to show anyone

3	2	5	3
8	4	1	9
8	2	6	3
9	3	7	1

Table 2. Prime 4-Square having 9 distinct digits.

how many and puts them in his pocket. Then he asks the Sand Counter to come back in the room and to say how many grains have been removed. The Verifier can then check. The Verifier can repeat this test until he is convinced that the Sand Counter is either incredibly lucky (zero-knowledge proofs are often probabilistic in nature) or is telling the truth. The Sand Counter is happy to answer these questions as long as the number of grains removed is tiny compared to the number in the bucket. This way he can be sure that the Verifier has learned how many grains there are.

References

[1] Hee-Kap Ahn, Siu-Wing Cheng, Otfried Cheong, Mordecai Golin, and Rene van Oostrum. "Competitive Facility Location: The Voronoi Game". Theoretical Computer Science, to appear. A preliminary version appears as "Competitive Facility Location along a Highway" in Proceedings of the 7th Annuual International Conference on Computing and Combinatorics (COCOON), 2001, pp. 237–246. See http://www.cs.ust.hk/~scheng/pub/pub.html.

[2] Otfried Cheong, Sariel Har-Peled, Nathan Linial, and Jir Matousek. "The One-Round Voronoi Game" Discrete and Computational Geometry, to appear. A preliminary version appears in Proceedings of the 18th Annual ACM Symposium on Computational Geometry, 2002, pp. 97–101. See http://www.win.tue.nl/~ocheong/Papers/.

[3] Sandor P. Fekete and Henk Meijer. "The one-round Voronoi game replayed" In Proceedings of the 8th Workshop on Algorithms and Data Structures, volume 2748 of Lecture Notes in Computer Science, 2003, pp. 150–161. See http://arXiv.org/abs/cs.CG/0305016.

[4] Oded Goldreich and Hugo Krawczyk. On the composition of Zero-Knowledge Proof Systems. SIAM Journal on Computing, vol. 25, no. 1, pp. 169-192, 1996.

[5] Atsuyuki Okabe, Barry Boots, Kokichi Sugihara. *Spatial Tessellations: Concepts and Applications of Voronoi Diagrams.* John Wiley & Son Ltd, New York, September 1992.

[6] Dennis Shasha. *The Puzzling Adventures of Dr. Ecco.* W. H. Freeman, New York, 1988. Dover, New York, 1997.

[7] Dennis Shasha. *Codes, Puzzles, and Conspiracy.* W. H. Freeman, New York, 1992.

[8] Dennis Shasha. *Dr. Ecco's Cyberpuzzles: 36 Puzzles for Hackers and Other Mathematical Detectives.* W. W. Norton, New York, 2002.

[9] In conversation, Andrei Broder suggested a clever solution in the case that the grains were all the same size. It involved a binary division of the bucket followed by a test involving adding a few grains or removing some to probe for lying. Once the Sand Counter had succeeded, one of the halves would be used in the next division.

The Complexity of Sliding-Block Puzzles and Plank Puzzles

Robert A. Hearn

Sliding-block puzzles have long fascinated aficionados of recreational mathematics. From Sam Loyd's infamous 14-15 puzzle to the latest whimsical variants such as Rush Hour^{TM}, these puzzles seem to offer a maximum of complexity for a minimum of space.

In the usual kind of sliding-block puzzle, one is given a box containing a set of rectangular pieces, and the goal is to slide the blocks around so that a particular piece winds up in a particular place. A popular example is Dad's Puzzle, shown in Figure 1; it takes 59 moves to slide the large square to the bottom left. What can be said about the difficulty of solving this kind of puzzle, in general? Martin Gardner devoted his February, 1964 Mathematical Games column to sliding-block puzzles. This is what he had to say [2]:

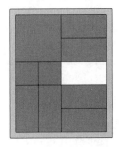

Figure 1. Dad's Puzzle.

Bob Hearn cowrote the Macintosh program ClarisWorks (now called AppleWorks). He is currently studying artificial intelligence at MIT.

These puzzles are very much in want of a theory. Short of trial and error, no one knows how to determine if a given state is obtainable from another given state, and if it is obtainable, no one knows how to find the minimum chain of moves for achieving the desired state.

Forty years later, we still do not have such a theory. It turns out there is a good reason for this: sliding-block puzzles have recently been shown to belong to a class of problems known as *PSPACE-complete*. These problems are thought to be even harder than their better-known counterparts, the NP-complete problems (such as the traveling salesman problem).

In 2002, Gary Flake and Eric Baum showed that Rush Hour is PSPACE-complete. Inspired by their work, Erik Demaine and I were able to show that ordinary sliding-block puzzles (without the lengthwise movement restrictions that Rush Hour imposes) are also PSPACE-complete.

Here I will sketch how to build computers out of sliding-block puzzles, one of the oldest staples of recreational mathematics. I will also introduce *plank puzzles*, one of the newest mathematical recreations, and show that they too possess this computational character.

Complexity Theory

Let's begin with the concept of PSPACE-completeness. Technically, a problem is called PSPACE-complete if it is equal in computational power to a particular mathematical model of computation (called "polynomial-space-bounded Turing machines"). Practically, this means that one can build computers out of elements of the problem, just as one can with wires and transistors.

But how can a sliding-block puzzle have computational power? That is what is so intriguing about these results! Of course, puzzles don't *do* anything on their own (except lure the unwary); they require users, and the users may slide the blocks around any which way they wish, without regard to any supposed computational sequence.

But a puzzle *can* correspond to a computer, in the following sense: if you give me a mathematical problem that such a computer can solve (say, by printing YES or NO), then I can give you back a sliding-block puzzle that has a solution if and only if the computer would print YES.

As a result of this computational property, it is natural to expect that sliding-block puzzles can be made that are very hard indeed: computers can solve difficult mathematical problems, and this difficulty can be translated directly into the difficulty of a puzzle. "Very hard" in this case means that we should not expect to do better than a brute-force search of all the possible move sequences.

Sliding-Block Logic Gates

To make all this a little more concrete, let's see how we might go about building computers out of sliding-block puzzles. Figure 2 shows how to make various logic gates and wiring elements from sliding-block puzzles.

Consider the AND mini-puzzle. Suppose the goal is to slide the upper protruding block into the box—how can we do this? The answer is that first the left block and the bottom block must both be slid out one unit. This will let the internal blocks slide to free up space for the upper block to slide in. (The light gray blocks are spacers, which don't ever need to move.) So, this puzzle does indeed have an AND-like property: both the left *and* the bottom blocks must slide out before the top one may slide in. Likewise, the OR mini-puzzle has an OR-like property: if either the left *or* the right block slides out, the internal blocks can be manipulated to allow the upper block to slide in.

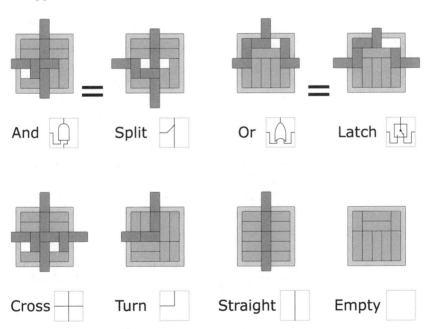

Figure 2. Sliding-block logic gates.

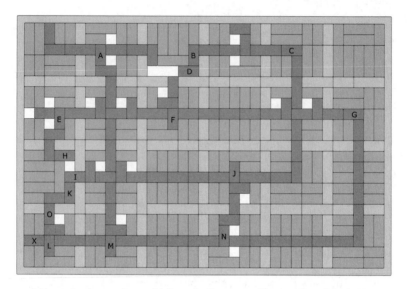

Figure 3. A puzzle made from logic gates. Try to move block X.

Just as real computers are made by wiring together logic gates, so we can make a sliding-block computer by connecting a lot of these mini-puzzles together, in a very large box. Figure 3 shows a 4 × 6 grid of these gates assembled into one large puzzle. (Can the reader see how to move block X?) The gates are arranged so that the "input" and "output" blocks are shared between adjoining gates. This is how signals flow throughout the puzzle.

Sometimes, wires have to cross over each other; the CROSS mini-puzzle accomplishes this: the left-right motion is independent from the up-down motion. Also, sometimes signals need to be split, and sent off in multiple directions. The SPLIT mini-puzzle does this; if the upper block slides out, then either or both of the bottom and left blocks can be made to slide in. But notice that this is really just an AND gate, used backwards!

That is one of the intriguing properties of this strange, nondeterministic kind of logic: signals can flow both forwards and backwards, and in fact they don't even need to respect normal notions of input and output! For example, block O in Figure 3 appears to be joining an AND input to an OR input. They key property, as it turns out, is that the AND and OR *constraints* are satisfied by any configuration of the puzzle, without regard to what would normally be considered an input or an output.

Perhaps the reader has noticed one omission from the menagerie of logic gadgets: there is no inverter, or NOT gate. Inverters are essential in ordinary digital logic, but they are not possible in this style of logic. To build an inverter, we would need a gate that enabled an output block to slide based on the presence, rather than the absence, of an input block. But a block may only enable another block to move by opening up a hole as it moves out of the way. And yet, it turns out that with this nondeterministic, constraint-based logic, all you need to build computers are AND and OR, and the ability to wire them together; you can get by without inverters.

The gates in Figure 2 are all made from 1×2 and 1×3 blocks (dominoes and trominoes). Can they be built using only dominoes? The answer is yes, but then they are much more complicated [3].

Sliding-Block Computers

The formal details of the PSPACE-completeness proof are beyond the scope of this essay, but the intuition that one can assemble a sliding-block computer from a collection of logic gates and wiring is the essential idea, and turns out to be valid. With a sufficiently large puzzle, we can arrange things so that the only way to solve the puzzle is to effectively perform a computation, by sliding the blocks following the sequence of logic gate activations a real computer would perform.

Flake and Baum showed explicitly how to build a kind of reversible computer using logic gates made from Rush Hour configurations, similar to the sliding-block gates presented here [1]. Their construction is quite ingenious and elegant.

Demaine and I took a slightly different approach. We started with a mathematical formulation of the underlying logic in terms of "constraint graphs," which turn out to be equivalent to the kinds of circuits shown here. Rather than explicitly build a computer, we showed how a problem (called Quantified Boolean Formulas) that is known to be PSPACE-complete can be translated into a constraint graph problem, and from there into a sliding-block puzzle. This is a common approach in computational complexity theory: by reducing problem A to problem B, one shows that B is at least as hard as A.

One of the subcircuits used in our proof is called a "universal quantifier" (Figure 4); this circuit is the basis for the puzzle in Figure 3. A long string of these circuits, connected together, effectively forces a particular computation to occur in order to solve the puzzle. Each extra universal quantifier also doubles the number of moves required—a string of n quantifiers takes on the order of 2^n moves to solve. This means that not only are there puzzles for which it is difficult to determine whether there is a

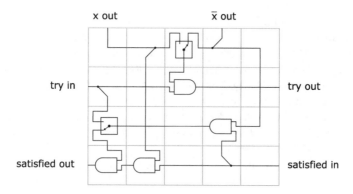

Figure 4. A "universal quantifier" circuit.

solution, but there are also puzzles that require a vast number of moves to actually solve.

Plank Puzzles

Sliding-block puzzles date from the 19th century. Now let's jump to the 21st century, for the latest in combinatorial challenges: *plank puzzles*, invented by UK maze enthusiast Andrea Gilbert. Like sliding-block puzzles, plank puzzles can hide enormous complexity behind a tame appearance.

The rules are simple. You have to cross a crocodile-infested swamp, using only wooden planks supported by tree stumps. You can pick up planks and put them down between other stumps, as long as they are exactly the right distance apart. You are not allowed to cross planks over each other, or over intervening stumps, and you can carry only *one* plank at a time.

A sample plank puzzle is shown in Figure 5. The first few moves of the solution are as follows: walk across the length-1 plank; pick it up; lay it down to the south; walk across it; pick it up again; lay it down to the east;

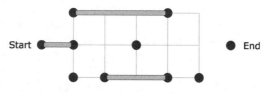

Figure 5. A plank puzzle.

walk across it again; pick it up again; walk across the length-2 plank; lay the length-1 plank down to the east... Can the reader see how to finish the sequence, and safely cross the swamp?

Many challenging plank puzzles are available on Ms. Gilbert's website, www.clickmazes.com, as playable Java applets. Plank puzzles are also available in physical form, from Binary Arts (called River Crossing^{TM}).

Plank-Puzzle Logic Gates

Just as with sliding-block puzzles, we can build nondeterministic, constraint-based computers out of plank puzzles. Thus, plank puzzles are also PSPACE-complete, and we should not be surprised that they can be very difficult.

The plank-puzzle AND and OR gates are shown in Figure 6. The length-2 planks serve as the input and output ports. To see how these gates work, consider the AND gate. Both of its input port planks (A and B) are present, and thus activated; therefore, you should be able to move its output port plank (C) outside the gate. Suppose you are standing at the left end of plank A. First walk across this plank, pick it up, and lay it down in front of you, to reach plank D. With D you can reach plank B. With B and D, you can reach C, and escape the gate. (Note that in sliding-block gates, signals propagate by blocks sliding *backwards* to fill holes, but in plank-puzzle gates, signals propagate by planks moving *forwards*.)

To complete the construction, we must have a way to wire these gates together into large puzzle circuits. It's all well and good to activate a single

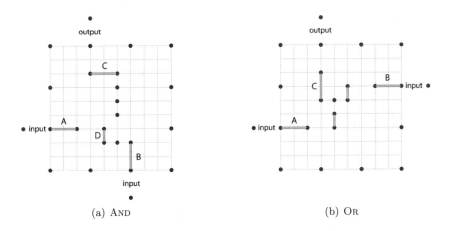

(a) AND   (b) OR

Figure 6. Plank-puzzle AND and OR gates.

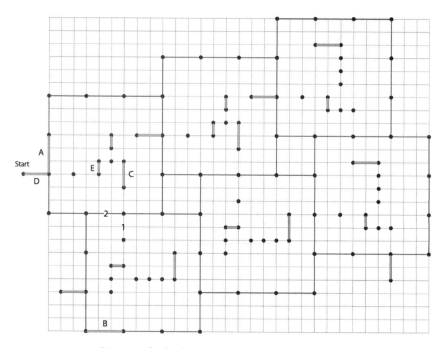

Figure 7. A plank puzzle made from logic gates.

AND gate, but when you've done that, you're stuck standing on its output plank—now what?

Figure 7 shows a puzzle made from six gates. For reference, the equivalent circuit is shown in Figure 8. The gates are arranged on a staggered grid, in order to make matching inputs and outputs line up. The port planks are shared between adjoining gates. Notice that two length-3 planks have been added to the puzzle. These are the key to moving around between the gates. If you are standing on one of these planks, you can walk along the edges of the gates, by repeatedly laying the plank in front of you, walking across it, then picking it up. This will let you get to any port of any of the gates. However, you can't get inside any of the gates using a length-3 plank, because there are no interior stumps exactly three grid units from a border stump.

Suppose you want to move plank C to position 1 (thus activating an OR output). This is what you do: first, walk plank A over to plank B. You can walk both of these planks together until B is in position 2, by alternately laying each plank in front of the other. Then walk A back to

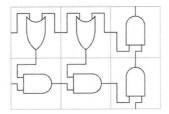

Figure 8. The equivalent circuit for Figure 7.

its starting position. Now, when you activate the OR, by using planks D and E, plank B is sitting there waiting for you at the gate exit. You can then use it and plank A to position yourself for the next gate activation. Can the reader see how to finish solving the puzzle?

As a result of this construction, everything we have said about sliding-block puzzles also applies to plank puzzles: it can be very difficult to determine whether there is a solution, and there are puzzles that take exponentially many moves to escape.

Conclusion

Puzzle PSPACE-completeness is not reserved for sliding-block puzzles and plank puzzles alone. The methods described here have been applied to several other kinds of puzzles and problems as well. The list so far includes Rush Hour, ordinary sliding-block puzzles, plank puzzles, Sokoban, hinged polygon dissections, and many related block-pushing problems. One might speculate that any sufficiently interesting motion-planning puzzle is PSPACE-complete, but there seems no hope of proving this in the abstract.

The lack of a successful theory of sliding-block puzzles, after all these years, could be seen as a mathematical failure. But to me, the reasons for the failure are inextricably linked to the very reason the puzzles are interesting: one can build all sorts of weird and wonderful gadgets with them. Furthermore, there is *still* a theoretical possibility for a successful theory! PSPACE-complete problems are indeed believed to be very hard, but nobody has yet proven that there is not some efficient algorithm that solves them all. An efficient method for solving sliding-block puzzles would yield the most important result in computer science of all time.

Answers

"Universal quantifier" puzzle (Figure 3). Only key moves are given. Slide blocks as far as possible. E left, D left, F up, G left, N right, J down, C left, I right, H left, K up, J left, B up, N up, F right, D right, A right, F up, G left, M up, L up, X right.

Small plank puzzle (Figure 5) conclusion. Pick up length-2, lay it north, pick up length-1, walk to far end of length-3, lay length-1 south, go back and pick up length-2, go back to far end of length-1, lay length-2 east, go back and pick up length-3, go back to end of length-2, lay length-3 east, escape.

Large plank puzzle (Figure 7). Figure 9 shows the puzzle after moving plank C as described. Before each step below, position A and B as needed. Move F to 3. Using C, G, and F, move H to 4. Using I and J, move K to 5. Using K, L, and H, move M to 6. Move N to 7. Using M, O, and N, move P to 8. Using Q, R, and P, move S to 9 and escape.

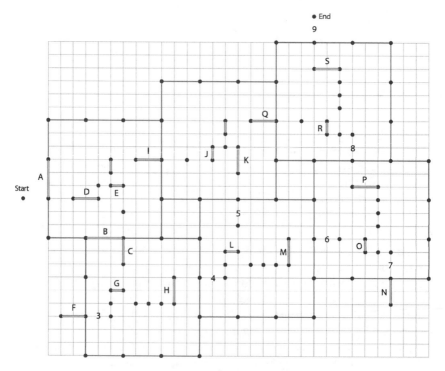

Figure 9. Figure 7 after activating an OR.

Acknowledgments

Thanks to Erik Demaine of MIT for pointing me towards the complexity problem for sliding-block puzzles, and for all the help in deriving the results. Thanks to Michael Albert of the University of Otago for introducing me to plank puzzles, and for help in designing the logic gates. Thanks to Andrea Gilbert for permission to reproduce one of her plank puzzles. Thanks to Martin Gardner for 25 years of Mathematical Games.

References

[1] Gary William Flake and Eric B. Baum. *Rush Hour* is PSPACE-complete, or "Why you should generously tip parking lot attendants". *Theoretical Computer Science*, 270(1–2):895–911, January 2002.

[2] Martin Gardner. The hypnotic fascination of sliding-block puzzles. *Scientific American*, 210:122–130, 1964. Also appears as "Sliding-Block Puzzles" in *Sixth Book of Mathematical Games from Scientific American*, W. H. Freeman, New York, 1971.

[3] Robert A. Hearn and Erik D. Demaine. The nondeterministic constraint logic model of computation: Reductions and applications. In *Proceedings of the 29th International Colloquium on Automata, Languages, and Programming*, volume 2380 of *Lecture Notes in Computer Science*, pages 401–413, 2002.

[4] Edward Hordern. *Sliding Piece Puzzles*. Oxford: Oxford University Press, 1986.

Hinged Dissections:
Swingers, Twisters, Flappers

Greg N. Frederickson

Introduction

A *geometric dissection* is a cutting of a geometric figure into pieces that can be rearranged to form another figure [3, 9]. When performed on two-dimensional figures, they are lovely demonstrations of the equivalence of area that have intrigued people over the ages. Dissections date back to Arabic-Islamic mathematicians a millennium ago and Greek mathematicians more than two millennia ago. They enjoyed great popularity as mathematical puzzles a century ago, in newspaper and magazine columns written by the American Sam Loyd and the Englishman Henry Ernest Dudeney [1, 10]. Loyd and Dudeney clearly established the goal of minimizing the number of pieces. Their puzzles engaged and entranced readers, especially when Dudeney introduced a novel variation in his 1907 book, *The Canterbury Puzzles* [2]. After presenting the remarkable 4-piece solution for the dissection of an equilateral triangle to a square, Dudeney wrote:

Greg Frederickson is a professor of computer science at Purdue University (West Lafayette, Indiana) and the author of two books on geometric dissections.

I add an illustration showing the puzzle in a rather curious practical form, as it was made in polished mahogany with brass hinges for use by certain audiences. It will be seen that the four pieces form a sort of chain, and that when they are closed up in one direction they form the triangle, and when closed in the other direction they form the square.

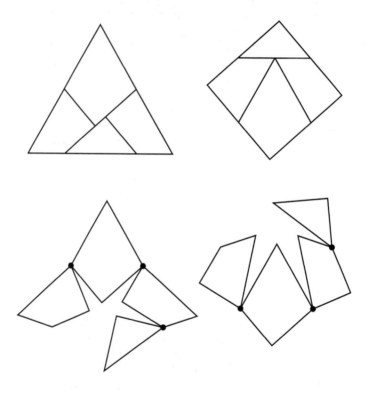

Figure 1. Swing-hinged dissection of a triangle to a square.

This *swing-hinged* model, in Figure 1, has seized the imagination of readers ever since. It has been described in at least a dozen other books in the intervening years, has been incorporated into the design of tables, and has even been adapted as a promotional item for a global banking group. When I wrote my first book on geometric dissections [3], I became captivated by hingeable dissections. I started to discover more and more hinged dissections, eventually filling up almost a whole book [4] with these striking inventions.

Recently, a second type of hinge has come into play. A *twist hinge* has a point of rotation on the interior of the line segment along which two pieces touch edge-to-edge. This allows one piece to be flipped over relative to the other, using rotation by 180° through the third dimension. Although twist hinges allow only limited movement, I have been able to find a relatively large number of twist-hinged dissections. Near the end of [4], there are an abundance, of which we will sample a few in the section, Twist-Hinged Dissections.

There is yet a third way to hinge two pieces: with a piano hinge. (The term refers to the long hinge that attaches the top of a grand piano to the box containing the strings.) To deal with this type of folding movement, the model of a two-dimensional figure must necessarily change, so that it consists of two *levels*, which we will assume to be of positive thickness. Two pieces that are *piano-hinged*, or *fold-hinged*, together can rotate from being next to each other on the same level to being on different levels with one on top of the other. Dissections in which the pieces "flap together" have appeared only a few times as curiosities. In [5] there are many more, of which we will inspect a few in the section, Piano-Hinged Dissections.

Swing-Hingeable Dissections

We can produce some swing-hinged dissections by using tessellations. Given a plane figure, a *tessellation of the plane* is a covering of the plane with copies of the figure without gaps and without overlap [7]. The technique of *superposing tessellations* [3, 9] takes two tessellations with the same pattern of repetition and overlays them so that the combined figure preserves this common pattern of repetition. If the points of intersection between line segments are at points of rotational symmetry, then we can induce a dissection that is hingeable. Harry Lindgren [8] gave a beautiful six-piece unhingeable dissection of a dodecagon to a square, which I have adapted to produce an eight-piece swing-hingeable dissection. The swing-hingeable dissection in Figures 2 and 3 is a bit different from the version in my book [3]. I discovered a way to cut and hinge the pieces so that two hinges are never adjacent to each other.

Another dissection technique is the strip technique [3, 9]: We cut a figure into pieces that form a *strip element* and then fit copies of this element together to form an infinite strip. We do the same for the second figure. Finally, we *crosspose* the two strips, positioning the strips so that the two boundaries of one strip cross a boundary of the other strip at points that are at a distance equal to the length of the element of the other strip.

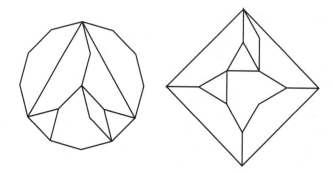

Figure 2. Swing-hingeable regular dodecagon to a square.

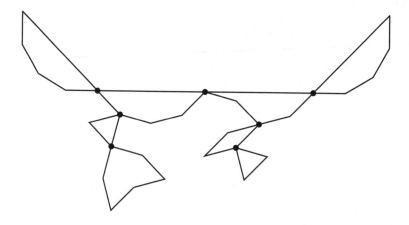

Figure 3. Swing-hinged pieces: dodecagon to square.

When we crosspose two strips, we force edges to cross only at points of rotational symmetry. Harry Lindgren [9] gave a seven-piece unhingeable dissection for a regular hexagon to a regular pentagon. I have found a ten-piece swing-hingeable dissection (Figs. 4 and 5).

Twist-Hinged Dissections

For a really neat example, let's return to the swing-hinged dissection of an equilateral triangle to a square in Figure 1. I use a triangle-swiping technique to give the seven-piece twist-hingeable dissection on the right in

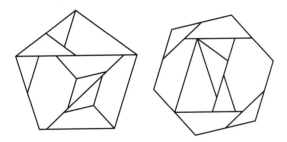

Figure 4. Swing-hingeable dissection: hexagon to pentagon.

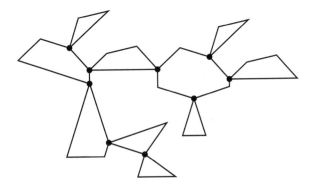

Figure 5. Swing-hinged pieces: hexagon to pentagon.

Figure 6. Each of the three right triangles results from borrowing isosceles triangles from two pieces attached by a hinge and then merging them together. Figure 7 shows intermediate steps in converting the triangle to the square.

In the twist-hinged world there is an amazing class of dissections. For any $p > 2$, there is a $(2p + 1)$-piece twist-hinged dissection of a regular polygon with $2p$ sides to a regular polygon with p sides. We see the example for a decagon and a pentagon ($p = 5$) in Figure 8.

We would expect a 13-piece dissection of the dodecagon to the hexagon in this family, but we can do better. Using techniques described in [4], plus one more discovered later, we can find a nine-piece twist-hingeable dissection in Figure 9. Eight of the pieces are cyclicly hinged.

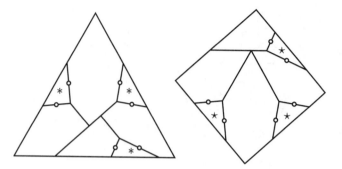

Figure 6. Twist-hinged dissection of a triangle to a square.

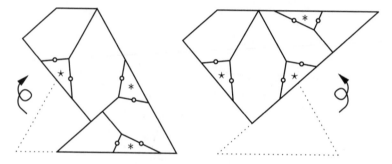

Figure 7. Intermediate configurations for the triangle to the square.

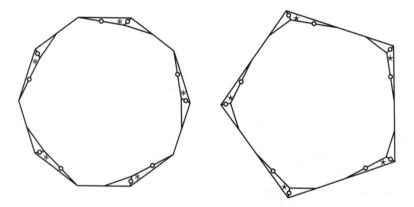

Figure 8. Twist-hinged dissection of a decagon to a pentagon.

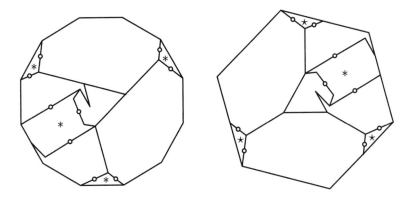

Figure 9. Twist-hinged dissection of a dodecagon to a hexagon.

Piano-Hinged Dissections ("Flappers")

Let's move on to dissections that will make you flip! (Or is it flap?) A pentomino [6] is a polygon produced by gluing five congruent squares edge-to-edge. Each of the twelve possible pentominoes has a piano-hinged dissection to a square. I present my eight-piece piano-hinged dissection of a W-pentomino to a square in Figures 10 and 11. I indicate a piano hinge that connects a piece on the top level to a piece on the bottom level by a line of dots next to the hinge line on each of the two levels. To indicate a piano hinge between two pieces on the same level, I use a true dotted line segment.

A dissection is *cyclicly hinged* if removing one of the hinges does not disconnect the pieces. In a *vertex-cyclic hinging*, four or more pieces touch at a vertex and each piece is hinged with its predecessor and successor on the cycle. If the angles that meet at the vertex sum to less than 360°, then the vertex-cyclic hinging is a *cap-cyclic hinging*. There are three examples of cap-cycles in Figures 10 and 11: Manipulating two of the pieces, say B and C, will cause all of the pieces to move simultaneously. If the angles that meet at the vertex sum to exactly 360°, then the vertex-cyclic hinging is a *flat-cyclic hinging*. If the angles that meet at the vertex sum to more than 360°, then the vertex-cyclic hinging is a *saddle-cyclic hinging*. I give examples for these other types of cyclic hingings in [5].

Our next dissection extends to a whole class of star polygons. For any whole numbers $p \geq 5$ and $q \geq 2$, with $q < p/2$, let $\{p/q\}$ be a star polygon with p points (vertices), where each point is connected to the qth points clockwise and counterclockwise from it. Lindgren [9] identified

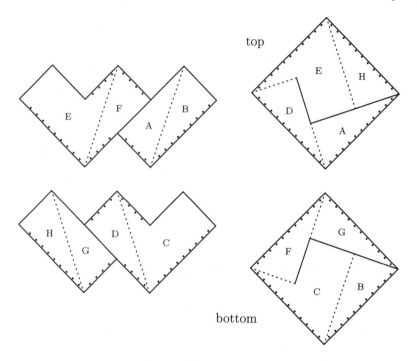

Figure 10. Folding dissection of a W-pentomino to a square.

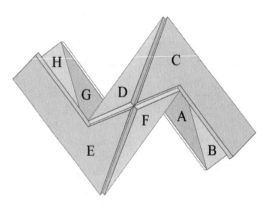

Figure 11. Projected view of a W-pentomino to a square.

$(4n + 2)$-piece swing-hingeable dissections of a $\{(4n + 2)/(2n - 1)\}$ to two $\{(2n + 1)/n\}$s. I have adapted them to be piano-hinged, obtaining an $(8n + 4)$-piece piano-hinged dissection. For $n = 2$, we get the case of two pentagrams to a $\{10/3\}$, in Figures 12 and 13.

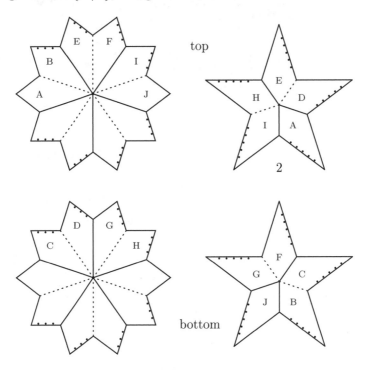

Figure 12. Folding dissection of two $\{5/2\}$s to a $\{10/3\}$.

There is a remarkable relationship between twist-hinged and piano-hinged dissections, namely that we can convert a twist-hinged dissection into one that is piano-hinged. I have identified two general techniques to perform this conversion. Given an n-piece twist-hinged dissection, I can find a $(4n - 3)$-piece piano-hinged dissection in which each twist hinge is simulated by three piano hinges.

If a piece in a twist-hinged dissection has more than one twist hinge incident to it, then we may be able to introduce just one piece that when flipped out of the way, avoids all obstructions of that piece with others. If we can do this for all of the pieces, then our conversion will just double the number of pieces. Thus I converted the twist-hinged dissection of a triangle to a square in Figure 6 to get the piano-hinged dissection of Figure 14.

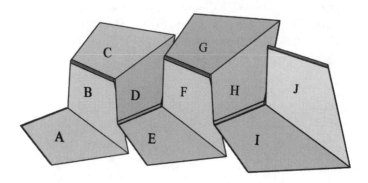

Figure 13. View of one assemblage for two $\{5/2\}$s to a $\{10/3\}$.

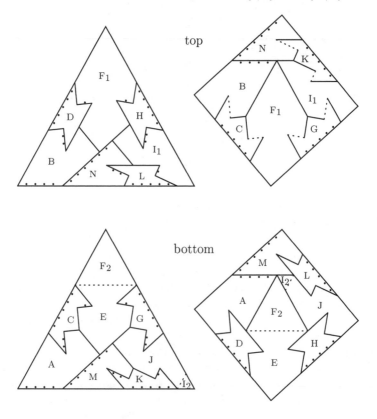

Figure 14. Piano-hinged dissection of a triangle to a square.

Again, there is a wealth of additional techniques and lovely piano-hinged dissections in [5], which will hopefully appear in print in due course.

References

[1] Henry E. Dudeney. Puzzles and prizes. Column in the *Weekly Dispatch*, April 19, 1896–March 27, 1904.

[2] Henry Ernest Dudeney. *The Canterbury Puzzles and Other Curious Problems*. W. Heinemann, London, 1907.

[3] Greg N. Frederickson. *Dissections Plane & Fancy*. Cambridge University Press, New York, 1997.

[4] Greg N. Frederickson. *Hinged Dissections: Swinging and Twisting*. Cambridge University Press, New York, 2002.

[5] Greg N. Frederickson. "Piano-Hinged Dissections: Time to Fold". Rough draft of a book in progress, 270 pages as of August, 2003.

[6] Solomon W. Golomb. *Polyominoes : Puzzles, Patterns, Problems, and Packings*. Princeton University Press, Princeton, 2nd edition, 1994.

[7] Branko Grünbaum and G. C. Shephard. *Tilings and Patterns*. W. H. Freeman and Company, New York, 1987.

[8] H. Lindgren. Geometric dissections. *Australian Mathematics Teacher*, 7:7–10, 1951.

[9] Harry Lindgren. *Geometric Dissections*. D. Van Nostrand Company, Princeton, New Jersey, 1964.

[10] Sam Loyd. Mental Gymnastics. Puzzle column in Sunday edition of *Philadelphia Inquirer*, October 23, 1898–1901.

Part VI

Braintools

The Burnside Di-Lemma: Combinatorics and Puzzle Symmetry

Nick Baxter

Introduction

Combinatorics, the mathematics of counting, provides invaluable tools for both puzzle solving and puzzle design. Solvers of mathematical and mechanical puzzles are often confronted with difficult issues of counting combinations, often complicated by symmetry. Similarly, puzzle designers may want to add elegance to their designs by incorporating symmetry and using sets of pieces that are somehow aesthetically pleasing in their completeness (such as the so-called English Selection[1]).

Nick Baxter is a mathematician and puzzle expert who loves to count.

[1] James Dalgety coined the phrase "English Selection," referring to a logically complete set of puzzle pieces. For example, the 12 planar Pentominoes qualify as an English selection; but for most puzzles, such the Soma Cube, Instant Insanity, Tangrams, Eternity, etc., this is not the case. Many times, whether or not a set is an English selection is a judgment call, and can be artificially contrived since the reference domain can be arbitrary constructed.

Conventional techniques are not always sufficient to solve some combinatorial problems, especially those where symmetry reduces the number of unique configurations. Fortunately, there is a particularly powerful, but relatively unknown tool for exactly this type of problem: the *Pólya-Burnside Lemma*. This paper will present this principle in common language and give specific examples of how it can be used.

Two-Color Cube Problem

Let's say that we want to determine the total number of possible two-color Instant Insanity cubes. In other words, how many different ways can you paint the faces of a cube using no more than two specific colors, say black and white?

At first blush, it may seem as if the answer is simply $2^6 = 64$, since the cube has six faces, each of which can be colored two ways. But that answer double counts certain colorings. For example, the coloring in which two opposite faces of the cube are black and the others white gets counted three times: once when the black faces are top and bottom, again when they're front and back, and once more when they're left and right. So 64 is really just an upper bound on the answer that we're looking for.

The trick is to organize the counting so that every possible configuration is counted *exactly once*. One approach is to first inventory the ways to partition the faces into at most two like-colored sets, and then later apply the colors to the patterns found:

- 6-0: Clearly, there's just one way to have six faces of one color and none of the other (we'll worry about which color it is later).

- 5-1: There is only one way to have just one face colored different from the other five (remember, we are ignoring rotations).

- 4-2: There are two ways to do this. One has the "minority" color on opposites faces of the cube; the other has it on two adjacent faces.

- 3-3: There are two ways to do this also. One has like-colored faces all meeting at a common vertex (with the remaining faces similarly oriented at the opposite vertex); the other has like-colored faces in a row.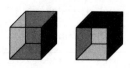

Now applying two colors to each of these patterns, we find 2, 2, 4, and 2 colorings respectively (because of symmetry, one must be careful not to count each of the 3-3 patterns twice), giving a total of 10, which is considerably less than our upper bound of 64.

For this problem, we had to be somewhat careful, but symmetry didn't cause too much trouble. The results would be the same if we "colored" the cube's faces with any two distinct markings, as long as both markings had 90° rotational symmetry, such as ☒ and ☑. But what if we used ⊟ instead? Now the trouble with symmetry becomes apparent: Do we have two patterns (⊟ and ⊡) or just one? Approaching the problem by inventorying the partitions of faces no longer works; and simply determining if two cubes are really the same becomes much more difficult to visualize. To better understand the difficulty, try solving this new problem; it will be discussed later in the examples.

These two problems can be extended, to four (or more) colors or orientations. With four options per face, there are $4^6 = 4{,}096$ total permutations. With so many possibilities, these problems move outside most puzzlers' scope of reliably determining inventories and symmetries by hand. One could write a search program, but such techniques are not necessarily of interest to mathematicians or non-programmers.

Pólya-Burnside Lemma

The good news is that the Pólya-Burnside Lemma[2] is the perfect tool for solving this type of problem. It's been known to mathematicians for over 100 years, but surprisingly I've found that many puzzlers, including those well-versed in recreational mathematics, are not familiar with this powerful technique. The rest of this paper presents the Pólya-Burnside Lemma without the jargon of combinatorial group theory and demonstrates how easy it is to use this powerful tool.

In the world of pure mathematics, any theorem must be precisely stated, including the conditions for when the theorem may be applied. Perhaps one reason why puzzlers don't know the usefulness of the lemma is that it is not always stated in terms that are easy to understand. For example, consider this version, taken from Eric Weisstein's MathWorld [11] (where it's called the Cauchy-Frobenius Lemma):

---

[2]Within mathematical circles, there has been plenty of discussion regarding the proper name and attribution, and it is probably best known recently as Burnside's Lemma. Neumann [8] gives an excellent history of this and a compelling case for the name Cauchy-Frobenius Lemma. However, I've chosen to recognize those that first applied the underlying principles to combinatorics and to use a name that appears to be most familiar with the intended audience.

Let J be a finite group and the image $R(J)$ be a representation
which is a homeomorphism of J into a permutation group $S(X)$,
where $S(X)$ is the group of all permutations of a set X. Define
the orbits of $R(J)$ as the equivalence classes under $x \sim y$, which
is true if there is some permutation p in $R(J)$ such that $p(x) = y$. Define the fixed points of p as the elements x of X for
which $p(x) = x$. Then the average number of fixed points of
permutations in $R(J)$ is equal to the number of orbits of $R(J)$.

Outside the world of group theory, this formulation doesn't help much.
Better is the concise statement from a text on combinatorial mathematics
by C. L. Liu [6]:

The number of equivalence classes into which a set S is divided
by the equivalence relation induced by a permutation group G
of S is given by

$$\frac{1}{|G|} \sum_{\pi \in G} \psi(\pi)$$

where $\psi(\pi)$ is the number of elements that are invariant under
the permutation π.

This formulation may be easy for mathematicians to understand, but
not for the rest of us. Let's start by translating the terminology that it
uses into language that a puzzler can understand.

S: The set of all possible variations of an object, with no considerations
for rotations and reflections. For painting a cube with four colors,
this set has 4^6 members.

Permutation: For puzzles, the permutations of interest are those that
take a physical object and reorient it so that it appears structurally
the same (ignoring coloring or other variations to be considered later).
For a cube, a permutation would be any way to pick it up and put
it back down in the same place. When doing so, any of the six
faces can be face down; then there are four ways to rotate the cube
so that the bottom face stays on the bottom. Thus there are 24
possible permutations of a cube. It is important to remember that a
permutation in this context is the action of transforming the cube to
a new position, not the position itself.

Permutation Group: This is the set of all possible permutations of the
physical objects under consideration. It forms a *group* in the formal
mathematical sense because the result of applying one permutation

and then another is again a permutation of the objects. The shorthand notation for the number of permutations in group G is $|G|$.

Equivalence Relation: This is the rule that determines whether or not two objects are the same. An "equivalence relation induced by a permutation group" is simply saying that two objects are the same if there is a permutation that transforms one into the other. So this is really what puzzlers mean when they say that two objects are "the same" or use phrases like "ignoring rotations and reflections."

Equivalence Class: When considering permutations of physical objects, this is a set of objects that are the same, ignoring rotations and reflections. When one is looking for the number of "unique shapes" or "unique colorings," we are really looking for the number of equivalence classes.

Invariant Element: This is an object that appears exactly the same before and after a permutation. For example, the first tile below is invariant when rotated $0°$, $90°$, $180°$, and $270°$; the second tile is invariant when rotated $0°$ and $180°$; and the third tile is invariant only when not rotated at all.

4way   2way   1way

The most important concept is that of invariant elements, because the Pólya-Burnside Lemma reduces all problems of symmetry to simply counting the number of invariant elements for each permutation. The key is that for many puzzles, this counting is significantly easier than any other equivalent problem-solving technique.

So it makes sense to first consider a *base object*, such as a cube, domino, tile, rectangle, etc., without any of the alternations or reorientations prescribed by the puzzle. Next, a *puzzle object* is a member of the set of all variations of the base object that satisfy the puzzle's constraints. Thus the typical puzzle will ask for the number of unique puzzle objects satisfying the given criteria.

Now we can restate the Pólya-Burnside Lemma using language that puzzlers can use.

Pólya-Burnside Lemma—Puzzlers' Version. The number of unique puzzle objects that are variations of a base object p is

$$\frac{1}{N}\Big(\psi(\pi_1) + \psi(\pi_2) + \cdots + \psi(\pi_N)\Big),$$

where $G = \{\pi_1, \pi_2, \ldots, \pi_N\}$ is the set of all physical permutations of p, and $\psi(\pi)$ is the number of invariant puzzle objects for the permutation π.

It may seem as if the Pólya-Burnside Lemma simply turns one counting problem into a multitude, since the number of permutations can itself be huge. But in theory—and in practice—permutations can be organized into families of similar operations, with the same $\psi(\pi)$ for each member in each family. (In the jargon of group theory, the families are called "conjugacy classes.") This often reduces a difficult counting problem to just a handful of easily solved counting problems.

All About the Cube

Let's return to the first example, coloring a cube with two colors, and see how this works. The base object p is just a normal cube, and G is the corresponding set of 24 permutations, so $n = 24$. The cube permutations can be grouped into families of similar operations as follows:

- I: Identity-no rotation (1 permutation)

- Q: 90° face rotation (6 permutations)

- H: 180° face rotation (3 permutations)

- D: 120° major diagonal rotation (8 permutations)

- E: 180° center-edge rotation (6 permutations)

For I, there are 2^6 ways to color the cube with two colors (ignoring rotations, reflections, or any other reorientation of the puzzle object). Since every object is invariant under the identity permutation, the total for this case is still 2^6.

For each of the six Q permutations, the top and bottom faces can be any color, since they stay in the same location during the rotation. For the four side faces, a Q permutation rotates one face to the next in a cycle of four. Thus they must all be the same color if a 90° rotation is to appear the same. This gives 2^3 invariant objects.

For each of the three H permutations, the top and bottom faces can again be any color. But unlike Q, an H permutation rotates each side face to the opposite face. Thus each opposite pair can be colored independently, and still leave the cube invariant after rotation. This gives 2^4 invariant objects.

For each of the eight D permutations, the cube is rotated around a major diagonal, and the two vertices it connects. The three faces touching each of those two vertices must be the same color. This gives 2^2 invariant objects.

For each of the six E permutations, the two faces adjacent to the edges that rotate must be the same. The top and bottom face must also be the same color. This gives 2^3 invariant objects.

In total, the Pólya-Burnside Lemma gives

$$\frac{1}{24}\Big(\psi(I) + 6\psi(Q) + 3\psi(H) + 8\psi(D) + 6\psi(E)\Big)$$
$$= \frac{1}{24}\Big(2^6 + 6 \cdot 2^3 + 3 \cdot 2^4 + 8 \cdot 2^2 + 6 \cdot 2^3\Big)$$

or 10 ways, agreeing with the previous solution. For this particular problem, using the Pólya-Burnside Lemma may have been more work than otherwise, but we should feel a lot better about not double-counting or missing any special cases.

Now for the magic—let's consider the same problem but with k colors instead of just two. Using the Pólya-Burnside Lemma, the analysis is almost identical to that of the two-color case:

$$\psi(I) = k^6, \quad \psi(Q) = k^3, \quad \psi(H) = k^4, \quad \psi(D) = k^2, \quad \psi(E) = k^3.$$

Thus the total number of unique colorings is

$$\frac{1}{24}\Big(k^6 + 3k^4 + 12k^3 + 8k^2\Big).$$

This result would have been very difficult[3] or impossible without the use of the Pólya-Burnside Lemma.

---

[3]This formula is different from the formula originally given by Gardner [3] in the chapter "The Calculus of Finite Differences" (but corrected in subsequent editions). That formula strangely worked only for cases $n = 1, 2, 3$, and 6—perhaps demonstrating the risk of relying solely on empirical results and finite differences for such problems.

Examples

We've shown that the Pólya-Burnside Lemma is general-purpose, relatively fast, and highly reliable. More importantly, it can help solve problems that would otherwise be next to impossible to solve. To demonstrate this power, I encourage you to imagine solving each of the following example problems using some other technique.

Second Cube Example

How many unique ways are there to paint a cube with a minus sign ($\boxminus$) on each face?

The two orientations of the minus sign ($\boxminus$ and $\boxed{|}$) behave like two distinct colors, and the analysis of invariant objects is the same as before—except for Q, the family of 90° rotations. In this case, the top and bottom faces must have 90° rotational symmetry if the cube is to be an invariant object. But since $\boxminus$ does not have this symmetry, the count for Q is zero. So instead, the total number of unique cubes is

$$\frac{1}{24}\left(2^6 + 0 + 3 \cdot 2^4 + 8 \cdot 2^2 + 6 \cdot 2^3\right) = 8.$$

Edge-Matching Tiles

How many different square edge-matching tiles are there using at most k colors?

Assuming that the tiles are one-sided, there are 4 permutations which can be grouped into three familiar categories: I, Q (2 cases), and H.

For I, there are k^4 invariant objects.

For Q, the four quadrants must all be the same color if the tile is to be invariant after a 90° rotation. Thus there are just k invariant objects. (Illustrated for $k = 2$.)

For H, a 180° rotation swaps pairs of opposite quadrants; thus there are k^2 invariant objects. (Illustrated for $k = 2$. Remember, when counting invariant objects, we ignore rotations; so the third and fourth figures are counted separately.)

The total number of colored tiles is

$$\frac{1}{4}(k^4 + 2 \cdot k + k^2).$$

Beveled Tiles[4]

In three dimensions, a square tile has a top face, a bottom face, and four side faces. In a beveled tile, each of the four side faces can have one of three styles: flat, angled in, or angled out. How many unique beveled tiles are possible?

A plain square tile is a flattened cube, and has only eight permutations: I, Q (2 cases), H (3 cases), and E (2 cases). Because not all the faces are square, there are actually two "flavors" of H that must be considered separately: H_1 is a 180° rotation about the center of the square face; H_2 includes the two 180° rotations about the center of a side face.

For I, each side face can have one of three styles, for 3^4 objects.

For Q, the tile can be rotated 90° about the center of the square face in either direction. To be invariant, a configuration must have the same style on all four side faces. Thus there are just 3 invariant objects.

For H_1, each pair of opposite side faces can have any of the three styles, giving 3^2 invariant objects.

For H_2, the two side faces being rotated must be the flat style in order to appear the same when turned upside down. The remaining two side faces must be paired, giving 3 invariant objects.

For E, the tile is rotated 180° about the axis connecting the centers of opposite short edges. One of the pair of side faces adjacent to such an edge can be any of the three styles, and will determine the style of the other. The same is true for the opposite edge, giving 3^2 invariant objects.

Thus, the total number of unique tiles is

$$\frac{1}{8}\left(3^4 + 2 \cdot 3 + 3^2 + 2 \cdot 3 + 2 \cdot 3^2\right) = 15.$$

---

[4]This problem was posed by Ed Pegg, Jr. at the 21st International Puzzle Party in Tokyo, August 2001, in preparation for a puzzle design he was considering.

Four Arrows[5]

How many different ways can you put six arrows on the faces of a cube?
(The arrows must be in one of four orthogonal orientations).

The analysis of invariant objects is the same as the k-color cube problem
(for $k = 4$), except that we must be more careful with the orientation of
the arrows.

Case I is the same as before: 4^6.

For Q, and H, the top and bottom faces must be invariant after 90° or
180° rotations, respectively. Since the arrow does not have such symmetry,
there are no invariant objects for these permutations.

For D, there are two cycles of three faces. Within each cycle, the arrow
on one face can be any orientation, forcing the orientation for each of the
other two (see faces shown in the above figure). Thus there are 4^2 invariant
objects.

For E, there are three pairs of faces that cycle. Similar to D, each cycle
can have four arrow orientations, giving 4^3 invariant objects.

The total number of cubes is

$$\frac{1}{24}\left(4^6 + 0 + 0 + 8 \cdot 4^2 + 6 \cdot 4^3\right) = 192.$$

In all our examples so far, the permutations have all corresponded in a
clear-cut way to some sort of physical movement. We finish with one more
example in which the notion of permutation is somewhat more subtle.

Numbered Slips

How many slips of paper are needed to individually print all (zero-filled)
n-digit numbers?

00691

This problem is tricky because we cannot use the expected
permutation group of $\{I, H\}$—a 180° rotation turns some numbers into invalid symbols, thus making it invalid as a permutation (see example).

Instead we must come up with an alternative permutation. C. L. Liu [6]
defines H' as a 180° rotation for those numbers containing only the digits

[5]This problem was most recently posed by Moscovich [7], problem #200.

$0, 1, 6, 8, 9$; otherwise it is the same as I. This is an improvement over H, since H' always gives a valid result. And since applying H' twice gives I, one can easily show that $\{I, H'\}$ is in fact a mathematically proper permutation group. With this, we can get back to looking at invariant objects.

For I, the number of invariant objects is 10^n.

For H', by definition, all numbers not made of only the symmetry digits $(0, 1, 6, 8, 9)$ are clearly invariant. This is $10^n - 5^n$ slips. For numbers using only symmetric digits, we must consider even and odd values of n separately.

For even n, each of the first $n/2$ digits can be any of five symmetric digits, forcing the selection of the last $n/2$ digits. This gives $5^{n/2}$ invariant objects.

For odd n, each of the first $(n-1)/2$ digits can be any of the five symmetric digits, forcing the selection of the last $(n-1)/2$ digits. The middle digit can be one of just three digits $(0, 1, 8)$ with $180°$ symmetry, giving a grand total of $3 \cdot 5^{(n-1)/2}$ invariant objects.

The total number of slips is

$$\frac{1}{2}\Big(\psi(I) + \psi(H')\Big) = \begin{cases} 10^n - (5^n - 5^{n/2})/2, & \text{for n even,} \\ 10^n - (5^n - 3 \cdot 5^{(n-1)/2})/2, & \text{for n odd.} \end{cases}$$

Beyond Pólya-Burnside

The Pólya-Burnside Lemma is actually just a special case of Pólya's Enumeration Theorem [9] (later generalized by de Bruijn [1]). If you know the cycle index of a permutation group, you can create a generating function that gives a *pattern inventory*, not just the total count.

For example, the cycle index for the permutation group of the six cube faces is

$$\frac{1}{24}\Big(x_1^6 + 6x_1^2x_4 + 3x_1^2x_2^2 + 8x_3^2 + 6x_2^3\Big).$$

For coloring the cube with two colors, represented by r and b, we simply substitute $(r^i + b^i)$ for x_i, giving

$$r^6 + r^5b + 2r^4b^2 + 2r^3b^3 + 2r^2b^4 + rb^5 + b^6,$$

which is the inventory of all possible two-color combinations. Looking back at the Two-Color Cube problem, we see that this corresponds exactly to the inventory that we found by hand, but derived without the risk of missing a case or double-counting.

For further reading, papers of interest not otherwise cited are Burnside [2], Golomb [4], Klass [5], and Read [10].

Acknowledgments

I'd like to acknowledge George Pólya for his inspirational lectures in my undergraduate combinatorics class—at 90 years old. I'd also like to thank Don Knuth and Stan Isaacs for their extensive and constructive comments on the draft, and to David Singmaster, whose inexhaustible enthusiasm rekindled my interest in this area.

References

[1] N. G. de Bruijn, Generalization of Pólya's Fundamental Theorem in Enumerative Combinatorial Analysis, *Proc. Koninkl. Nederl. Akad. Wetenschap.* A 62, (Indag. Math., 21), pp. 59-79, 1956.

[2] W. Burnside, *Theory of Groups of Finite Order*, Cambridge University Press, Cambridge, 1911.

[3] Martin Gardner, *Martin Gardner's New Mathematical Diversions from Scientific American*, Simon and Schuster, New York, 1966 (First Printing), pp. 234-246.

[4] Solomon W. Golomb, *Polyominoes*, Princeton University Press, Princeton NJ, 1994 (revised edition), pp. 43-69.

[5] Michael J. Klass, A Generalization of Burnside's Combinatorial Lemma, *Journal of Combinatorial Theory Ser. A*, 20:273-278, 1976.

[6] C. L. Liu, *Introduction to Combinatorial Mathematics*, McGraw-Hill, New York, 1968.

[7] Ivan Moscovich, *1000 PlayThinks*, Workman Publications, New York, 2001.

[8] Peter M. Neumann, A Lemma that is not Burnside's, *Math. Scientist*, 4:133-141, 1979.

[9] G. Pólya, Kombinatorishe Anzahlbestimmungen fr Gruppen, Grphen und chemische Verbindugen, Acta Math. 68:145-254, 1937.

[10] R. C. Read, Pólya's Theorem and Its Progeny, *Mathematics Magazine*, 60(5):275-282, Dec. 1987.

[11] Eric Weisstein, World of Mathematics, http://mathworld.wolfram.com/.

Designing Puzzles with a Computer

Bill Cutler

From an early age I have been interested in both puzzles and computers. As a computer programmer, I want to write a program that can solve a puzzle. As a puzzle designer, I want to design a puzzle that will be difficult to solve with a computer. On rare occasions, I can combine both—use a computer program to design a puzzle that will be difficult or impossible to solve with a computer. But most of the time, the challenge is to get the computer program to take the next step—to solve the next kind of puzzle that comes up, or to write a program that uses a completely different approach to solve a puzzle.

Early Puzzle Programs

The first puzzle program I ever wrote was never run on a computer. The year was 1965; I was an undergraduate student at Brown University, and the puzzle was a two-dimensional packing puzzle with ten black, red, and

Bill Cutler was an avid reader of the early Martin Gardner *Recreational Mathematics* columns. He works as a computer programmer at EDS and designs and analyzes puzzles for fun.

Figure 1. The Ten-Yen puzzle.

white pieces called Ten-Yen. (See Figure 1.) The program was supposed to find all solutions in which no two pieces with the same color touch each other. It goes without saying that computers were not as readily available in 1965 as they are now.

The first puzzle program I ever ran on a computer didn't fare much better. In 1970 I bought a nice wooden model of the pentominoes from Stewart Coffin. The problem was to pack them into a $3 \times 4 \times 5$ box. I decided I would try to solve it with a computer. The program just sat there without doing anything—I am sure it was stuck in a loop. I gave up and managed to find a solution by hand that afternoon. Three years earlier, J. C. Bouwkamp had finished a computer project that found 3,940 solutions to the same puzzle—this had taken him several years [1]. The current version of my general box-packing solver takes four minutes on a 2 GHz Pentium 4 processor to do the job. This says more about the increasing power of computers then it does about the programmers involved.

In 1974 I wrote the first program dealing with 6-piece burrs. This first one dealt only with solid, notchable 6-piece burrs, and the logic was very specific to these puzzles. Later, I expanded the program to deal with solid, unnotchable 6-piece burrs. Dealing with "holey" 6-piece burrs, however, would require a program that dealt in a very general way with interlocking puzzles.

General Disassembly Program

In the early 1980s, I wrote the first general disassembly program, which I called GENDA. GENDA can be used to analyze interlocking puzzles in which the underlying construction is an orthogonal grid of cubes all the same size. The program has the following restrictions:

- The pieces can only be moved in one of the three orthogonal directions.

- The distance a piece is moved must be an integer multiple of the cube size.

- Pieces can be moved individually or in groups of two or more pieces.

I then combined this with another program which determined all possible ways six pieces could fit together to make a 6-piece burr. Each of these "assemblies" was then submitted to GENDA to see if it could be physically constructed. The resulting program, BURR6, is a powerful tool for analyzing 6-piece burrs.

Stewart Coffin had designed a couple of "holey" 6-piece burrs in the early 1980s, and challenged his puzzle customers to submit their own designs. He would then make models of the design he judged to be the best. I used BURR6 to design a 6-piece level-5 burr, meaning that five moves must be made before the first piece can be removed. This puzzle, which I called Bill's Baffling Burr, had 24 assemblies but only one solution. (See Figure 2.) Stewart declared my entry the winner of the contest. Critics will

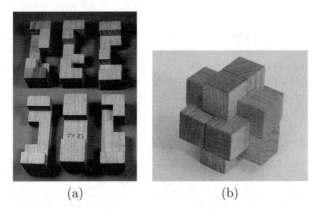

(a)   (b)

Figure 2. Bill's Baffling Burr, (a) in pieces and (b) assembled.

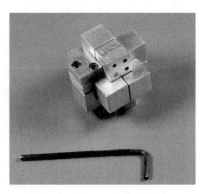

Figure 3. The aluminum Bill's Baffling Burr with Allen wrench.

point out that mine was the only entry, and that it was two years too late. Bill's Baffling Burr was written up in the Computer Recreations column of *Scientific American*, but one of the pieces had two extra cubes in it, and so the puzzle was impossible as printed [3]. The pieces as printed could be used to create two different assemblies, but neither one can be physically put together. I received a model from one reader made with aluminum. (See Figure 3.) You need an Allen wrench to take it apart. The reader thought that if he managed to get the pieces together somehow, then he would be able to see how it came apart.

From 1987 to 1990, I ran specialized versions of GENDA and BURR6 to analyze all 35.5 billion different 6-piece burr assemblies. I had a lot of help from people who ran the programs on their computers when they were not doing something else. In particular, Harry Nelson at Lawrence Livermore Labs did about a third of the analysis on the Cray computers that he administered. The main goal of the analysis was to find high-level burrs. In that respect, the project could be viewed as a failure, since the highest-level burr (level 12) found by the computer programs had already been discovered manually. The puzzle is called Love's Dozen and was designed by Bruce Love of New Zealand [2].

Twist Moves

I knew when I wrote the disassembly program that it had definite limitations. It cannot handle twist moves. Consider the construction shown in Figure 4, which I call the three-dimensional weave. It is so loose that you would have difficulty picking it up without it falling apart. But the com-

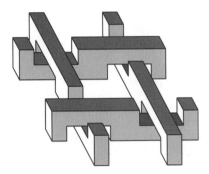

Figure 4. Three-dimensional weave, made from four pieces of the same shape. The computer is unable to disassemble the pieces.

puter can't figure out how to take it apart. It will find a lot of movement between the pieces, but you cannot separate them using linear moves in the three orthogonal directions.

One of Stewart Coffin's early puzzle designs, Convolution, features a twist move. Convolution is just a dissection of a $4 \times 4 \times 4$ cube. (See Figure 5(a).) It looks like the sort of puzzle that could be easily solved with a computer, but that is not the case. The first two pieces come out easily, but the next one is too tough for the computer—it starts out by making a simple linear move, but then it must be twisted. When you are making the twisting move, you can feel that there is some jamming going on with the pieces, but not nearly enough to stop you from making the twist. This twist is what I call an illegal twist. (See Figure 5(b).)

(a)   (b)

Figure 5. (a) Stewart Coffin's Convolution; (b) The twist move.

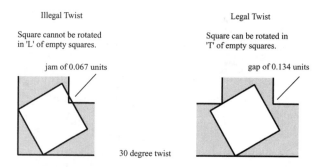

Figure 6. An illegal twist and a legal twist.

Legal and Illegal Twists

A legal twist is one that can be made even if the pieces are cut from a completely rigid material and are cut with no tolerance. Some illegal twists should be obvious to the puzzler. For example, in Figure 6 on the left—it is clearly illegal to try to rotate a square inside a hole which is an L-shaped hole of equal size squares. If such a construction is made from two pieces of wood, you should not be able to make the twist. However, with burr puzzles, it is more likely that this opening is made by two or three different pieces of the puzzle, which all have some movement between them, and the cumulative effect may result in the move being possible to make.

Is there a 6-piece burr which requires a fully legal twist to take it apart? At first I thought not, as there are very few pieces and not much room to "work around." But in 1988, during the 6-piece burr analysis, the computer discovered an assembly called Computer's Choice 5-Hole, which had the illegal twist on the left. I realized that with some slight modification, the twist could be made legal, and the piece shapes could be redesigned so that the twist was required to get the puzzle apart. In the legal twist on the right of Figure 6, you can twist a square in a T of similar size square holes.

Thus was born the Programmer's Nightmare. (See Figure 7.) I used a computer program to analyze thousands of 6-piece burrs which had the right pieces to implement this twist move, searching for one which had no "normal" solutions. The program found two sets of pieces that had only this twist move solution.

This is the first case of using a computer to design a puzzle that can't be solved by a computer. Even with the computer programs that I have now, I would have a difficult time finding the solution to this puzzle. My programs would determine that there were 102 assemblies, but no solutions.

Figure 7. The Programmer's Nightmare.

I would then have to manually go through these 102 assemblies, searching for one which allowed for some kind of twist move that the computer could not detect.

Box-Packing Program

In 1990 I wrote the first version of my general box-packing program BCP-BOX. Since my first ill-fated attempt at solving the 3 × 4 × 5 pentomino packing problem, I had written many programs that were successful in using "back-tracking" algorithms to find solutions to various packing puzzles. I realized it was time to write one program that could handle a variety of situations. Over the years, the program has been expanded to include a variety of types of packing problems. For example, in two-dimensions it can handle grids made out of squares, checkerboard squares, right isosceles triangles, hexagons, equilateral triangles, hexagons and triangles, squares and triangles, and drafters (30-60-90 triangles); and in three-dimensions it can handle grids made out of cubes, checkerboard cubes, rhombic dodecahedra, truncated octahedra, and triangular prisms.

After writing program BOX, my interests turned toward creating box-packing puzzles that my program couldn't solve. One of the earliest results of this is the Bermuda Hexagon. My original idea was to create a puzzle using triangular prisms in which some of the prisms on the inside of the puzzle were not in the positions one might expect. At that time, program BOX could not handle triangular prisms, and so this was added to the program.

(a)   (b)

Figure 8. Bill's Checkerbox, (a) in pieces and (b) assembled.

This idea of using unexpected grid patterns was expanded to include "checkerboard" puzzles. I thought of designing a puzzle using checkerboard cubes in which the color pattern was not consistent on the inside of the solved puzzle. I wanted the pieces to look like normal checkerboard pieces when not put together, so that the user would not suspect anything was different. This would require that any piece that was put in the "wrong way" would have to be completely inside the assembled puzzle.

The problem of packing the pentominoes into a 3×4×5 box seemed ideal for this purpose. The inside is of size 1×2×3, and only two of the pentominoes can fit into this space—the P-pentomino and the U-pentomino. There was an additional feature that came into play: in a normal packing of checkerboard pentominoes, there would always be exactly 30 cubes of each color. With exactly one of the pentominoes colored "wrong," there would then be 31 cubes of one color and 29 cubes of the other color. An observant puzzle-solver might count up the cubes of each color and realize that something is wrong. Thinking "outside the box" on the part of the potential solver should be rewarded!

I did a computer analysis of the 3,940 solutions to determine a coloring of the pieces which would have a unique solution. There were about a dozen such colorings to choose from—the one I picked is called Bill's Checkerbox. (See Figure 8.)

I designed two additional puzzles using variants of the "messed-up grid" idea—The Squash(ed) Box and the Splitting Headache, which is one of my most popular designs.

A few packing puzzles do not follow a consistent grid. These create a whole new problem for the computer programmer. Ed Pegg's puzzle Kites & Bricks is one such puzzle. (See Figure 9.) It uses only three different pieces, but they do not fill space in a regular fashion. This puzzle is not

Figure 9. Ed Pegg's Kites & Bricks.

solvable using my more standard box-packing program, but I was able to solve it using a different program based on joining polygons together.

Irregular Two-Dimensional Packings with Holes

OK, the gloves are off with these final designs. No regular grids; the pieces may or may not have regular shapes; and there are extra spaces in the tray in which they are packed. Even if the pieces followed a grid when they were packed, it would be difficult to solve because the most efficient algorithms require all spaces to be filled. However, if the pieces don't follow a grid, or they follow some grid, but you aren't sure which one, then trying to solve with a computer program is a whole new ball game.

Take Stewart Coffin's design the Engelberg Square. In part (a) of Figure 10 is the unsolved puzzle—six pieces with a total of 25 squares. The truncated corners look like an artistic effect, but we see that is far from the case. A 5 × 5 square of these octagons will fit nicely in the square box as in Figure 10(b), but this cannot be done—a computer program can easily verify this. One could also fit a pattern of 25 octagons in at a 45-degree angle as in Figure 10(c), but this also cannot be done—also verifiable with a computer program. The solution is a combination of part of each pattern. (See Figure 10(d).) Even if you know that you are looking for a solution like this, it would still be very messy to solve with a computer program.

(a)   (b)

(c)   (d)

Figure 10. The Engelberg Square. (a) The pieces; (b-c) Trying a "normal" solution.; (d) The solution.

However, from the puzzle designer's point of view, a computer program can be very useful in ruling out the unwanted "normal" solutions. I don't believe Stewart used a program for this particular puzzle, but he has started using programs for this purpose with some of his new designs.

I'll close with another Stewart Coffin design. (See Figure 11.) Again we have simple pieces—six pentomino pieces to fit into a rectangular box. Again, you can't fit the pieces into the box in a simple way. The solution, shown in Figure 11(b), is very elegant—again a mix of regular patterns that will not be found by a computer. It has the nice feature of also being pleasingly symmetric.

Unfortunately, the large percentage of empty space in the puzzle increases the likelihood of some other, unintended solutions; see, for example, parts (c) and (d) of Figure 11. How can we keep this from happening? It would be nice to have a program to check for such unwanted solutions, but how? When you think about it, the number of truly different ways

(a)   (b)

(c)   (d)

Figure 11. Stewart Coffin's design #175. (a) Trying a "normal" solution; (b) The elegant solution; (c-d) Two not-so-elegant solutions. (See Color Plate V.)

you could try and place these pieces in the tray is relatively small for a computer. The brain is quick at checking out these possibilities—surely you could program a computer to do it, but how?

I have started to write a computer program which I hope can be used to solve puzzles such as the last two. It is too early to tell whether this program has a chance of being successful.

I am reminded of my favorite definition of Artificial Intelligence—getting a computer to do something that you didn't think it was possible to get a computer to do. And I can hear Raymond Smullyan finishing off with the conclusion that there is no such thing as an Artificial Intelligence computer program.

References

[1] C. J. Bouwkamp, *Catalogue of solutions of the rectangular* $3 \times 4 \times 5$ *solid pentomino problem*, Technological University Press, Eindhoven, July, 1967.

[2] Bill Cutler, "A Computer Analysis of All 6-Piece Burrs", Available from http://www.billcutlerpuzzles.com, 1994.

[3] A. K. Dewdney, "Computer Recreations", *Scientific American*, pp. 16-27, October, 1985.

Origami Approximate Geometric Constructions

Robert J. Lang

Introduction

Compass-and-straightedge geometric constructions are familiar to most students from high-school geometry. Nowadays, they are viewed by most as a quaint curiosity of no more than academic interest. To the ancient Greeks and Egyptians, however, geometric constructions were useful tools, and for some, everyday tools, used for construction and surveying, among other activities.

The classical rules of compass-and-straightedge allow a single compass to strike arcs and transfer distances and a single unmarked straightedge to draw straight lines. The two may not be used in combination, for example, holding the compass against the straightedge to effectively mark the latter. However, there are many variations on the general theme of geometric constructions that include the use of marked rules and tools other than compasses for the construction of geometric figures.

Robert J. Lang is an artist and the author of 8 books on origami, as well as a contributor to the field of origami mathematics.

One of the more interesting variations is the use of a folded sheet of paper for geometric construction. The act of folding a sheet of paper and flattening the result creates a crease that is a straight line, while the intersection of two creases defines a point. Thus, folding alone can be used to create both points and lines, and consequently, many other geometric figures.

Like compass-and-straightedge constructions, folded-paper constructions are both academically interesting and practically useful. The practical application arises within *origami*, the art of folding uncut sheets of paper into interesting and beautiful shapes. However, many origami designs—even quite simple ones—require that one create the initial folds at particular locations on the square: dividing it into thirds or twelfths, for example. More complex figures require the construction of points that are defined by the solution of sets of coupled algebraic equations. While one can certainly compute, measure, and mark these points, there is an aesthetic appeal to creating these key points, known as reference points, purely by folding.

Thus, within origami, there is a practical interest in devising folding sequences for particular proportions that overlaps with the mathematical field of geometric constructions. There is now a growing body of work within both the origami and mathematical literature devoted to the field of of geometric constructions using folding alone. While much of the work in this field has focused upon exact geometric constructions, in recent years the specialty of approximate constructions has seen more work. Approximate constructions are of interest for their mathematical properties and for their practical application within origami. There is a natural limit to the precision of human folding, and in many cases, an efficient folding sequence for an approximation of a given point can provide superior accuracy in practice than a longer "exact" solution.

Preliminaries and Definitions

Origami, like geometric constructions, has many variations. In the most common version, one starts with an unmarked square sheet of paper. Only folding is allowed: no cutting. The goal of origami construction is to precisely locate one or more points on the paper, often around the edges of the sheet, but also possibly in the interior. These points, known as *reference points*, are then used to define the remaining folds that shape the final object. The process of folding the model creates new reference points along the way, which are generated as intersections of creases with one another or with the folded edge. In an ideal origami *folding sequence*—a step-by-

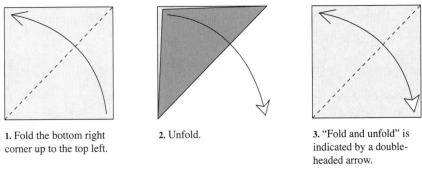

1. Fold the bottom right   2. Unfold.   3. "Fold and unfold" is
corner up to the top left.   indicated by a double-
                                                          headed arrow.

Figure 1. The sequence for folding a square in half diagonally.

step series of origami instructions—each fold action is precisely defined by aligning combinations of features of the paper, where those features might be points, edges, crease lines, or intersections of same.

Two examples of creating such alignments are shown in Figures 1 and 2. Figure 1 illustrates folding a sheet of paper in half along its diagonal. The fold is defined by bringing one corner to the opposite corner and flattening the paper. When the paper is flattened, a crease is formed that (if the paper was truly square) connects the other two corners.

As a shorthand notation, the two steps of folding and unfolding are commonly indicated by a single double-headed arrow as in the third step of Figure 1.

Figure 2 illustrates another way of folding the paper in half ("book-wise"). This fold can be defined in three distinct, but equivalent ways:

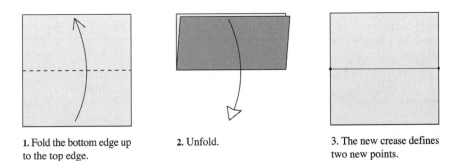

1. Fold the bottom edge up   2. Unfold.   3. The new crease defines
to the top edge.   two new points.

Figure 2. The sequence for folding a square in half bookwise.

1. Fold the bottom left corner up to the top left corner.

2. Fold the bottom right corner up to the top right corner.

3. Fold the bottom edge up to be aligned with the top edge.

For a square, these three methods are equivalent. However, if you start with slightly skew paper (a parallelogram rather than a square), you will get slightly different results from the three.

In both cases, if you unfold the paper back to the original square, you will find that you have created a new crease on the paper. For the sequence of Figure 2, you will also have now defined two new points: the midpoints of the two sides. Each point is precisely defined by the intersection of the crease with a raw edge of the paper.

These two sequences also illustrate the rules that we will adopt for origami geometric constructions. The goal of origami geometric constructions is to define one or more points or lines within a square that have a geometric specification (e.g., lines that bisect or trisect angles) or that have a quantitative definition (e.g., a point 1/3 of the way along an edge). We assume the following rules:

1. All lines are defined by either the edge of the square or a crease on the paper.

2. All points are defined by the intersection of two lines.

3. All folds must be uniquely defined by aligning combinations of points and lines.

4. A crease is formed by making a single fold, flattening the result, and (optionally) unfolding.

Rule (4), in particular, is fairly restrictive; it says that folds must be made *one at a time*. By contrast, all but the simplest origami figures include steps in which multiple folds occur simultaneously.

Binary Approximation

One of the simplest techniques to approximate a given point is to approximate its x and y coordinates separately by bisection. Given a point within a square, one can repeatedly fold two vertical lines together on either side of the x-coordinate, starting with the two edges of the square. Similarly,

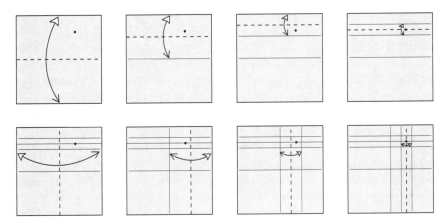

Figure 3. Successive approximation of a point by bisection in the x and y directions.

one can converge on the y-coordinate by folding together successive pairs of horizontal lines. This process is illustrated in Figure 3.

This process converges rapidly. With each fold, the maximum error in the x- or y-coordinate is halved, and so it is clear that one can construct an arbitrarily good approximation to any given point within the square by this method.

A different method for constructing the folding sequence for a given fraction of the side of the square was presented by Brunton [1] and elaborated upon by Lang [2]. If one writes the number to be approximated in binary notation, the binary expansion provides the folding prescription via the following rule:

> To mark off a distance equal to a binary fraction by folding, write down its binary expansion to k digits.
>
> Beginning from the *right* side of the fraction (the least significant digit): for the first digit (which is always a 1 because you can drop any trailing zeros) fold the top down to the bottom and unfold.
>
> For each remaining digit, if it is a 1, fold the top of the paper to the previous crease and unfold; if it is a 0, fold the bottom of the paper to the previous crease and unfold.

An example will make this clear. For $k = 5$, the fraction $29/37$ has the truncated binary expansion

$$29/37 \approx .11001_2. \tag{1}$$

0.11001

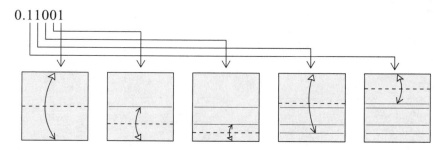

Figure 4. Folding sequence for 29/37, approximated as 0.11001_2. Each crease becomes the reference for the fold in the subsequent step with the last crease giving the desired approximation.

This gives the folding sequence shown in Figure 4.

By comparing this algorithm with the expanded formula for a binary fraction as a nested series, you can see how the folding algorithm works.

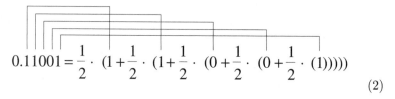

$$0.11001 = \frac{1}{2} \cdot (1 + \frac{1}{2} \cdot (1 + \frac{1}{2} \cdot (0 + \frac{1}{2} \cdot (0 + \frac{1}{2} \cdot (1)))))$$

$$(2)$$

To evaluate this form, you start at the innermost number in the expression (the terminal "1") and work your way back to the left and out of the nested parentheses. If we write the fraction this way, it becomes a series of nested operations where each operation is either:

(a) Add 0 and multiply by 1/2, or

(b) Add 1 and multiply by 1/2.

Now let's look at the origami folding sequence in the recipe above. If we have a square with a crease located a distance r from the bottom and fold the bottom of the square up and unfold, the new crease is made a distance $(1/2)r$ from the bottom, as shown in Figure 5. If instead, we fold the top of the square down to the crease and unfold, the new crease is made a distance $(1/2)(1 + r)$ from the bottom. Thus, folding the bottom up or top down is equivalent to performing operations (a) or (b), respectively.

Since any binary fraction can be written as a nested sequence of the two operations (a) and (b) and the two folding steps shown in Figure 5 implement these two operations, it follows that any proportion can be folded

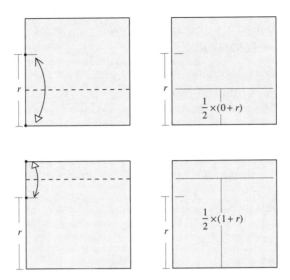

Figure 5. (Top) Folding the bottom edge up to a crease r gives a new crease $(r/2)$ from the bottom. (Bottom) Folding the top edge down to a crease r gives a new crease $((1 + r)/2)$ from the bottom.

to arbitrary accuracy from its binary expansion. The folding sequence derived from the binary expansion is usually easier to fold than the bisection sequence because every fold aligns the paper boundary to an internal line, rather than requiring the alignment of two internal lines as in Figure 3.

Accuracy

An important consideration in any approximation method is the tradeoff between the accuracy of the approximation and the number of folds that must be made. We can quantify the accuracy of a particular construction by computing the Euclidean distance between the desired point—called the *target point*—and the constructed approximate point, a quantity which we call the *error* of the approximation. We define the number of folds necessary to construct a given point as the *rank* of the point. An approximation scheme can then be quantified by the tradeoff between the rank of the constructed point and its error.

For any specific target point, it is usually a relatively simple process to compute the error of a given approximation. We can also quantify the accuracy of a general scheme over a collection of target points, for example, the uniform distribution over the unit square. A scheme can be characterized in several ways: we can compute the worst-case error; the mean error (assuming a uniform distribution of target points); or something else. It turns out that both worst-case and mean error are rather difficult to compute for any but the simplest of schemes. However, a useful alternative is to calculate the number of constructible points of a given rank—often a much easier problem—from which a useful estimate of the mean error may be calculated.

Suppose, for example, that a folding scheme allows the construction of N distinct points within the unit square. Then it can be shown that, if the points were uniformly distributed within the square, then as N becomes large, the mean error approaches the limit

$$\langle \varepsilon \rangle \sim \frac{1}{2\sqrt{N}}. \tag{3}$$

In fact, it is relatively uncommon that the constructible points from an approximation scheme are uniformly distributed; nevertheless, equation (3) turns out to be reasonably accurate for a variety of approximation schemes.

For the binary folding method, it is relatively simple to analyze the number of constructible points of a given rank. We first consider the one-dimensional problem, that is, approximating a single value with k folds. Given the one-to-one correspondence between a k-fold sequence and a k-digit binary expansion discussed above, it is clear that, in one dimension, exactly k folds allow us to construct all possible binary fractions in the range [0,1] that have exactly k digits in their binary expansion (not including trailing zeroes).

We denote the number of points constructible with exactly k folds (that is, the number of points of rank k) by $n^{(1)}(k)$. Then $n^{(1)}(0) = 2$, corresponding to the two initial values of 0 and 1. One fold constructs the value 1/2, so $n^{(1)}(1) = 1$; two folds constructs the values 1/4 and 3/4, so $n^{(1)}(2) = 2$; three folds constructs 1/8, 3/8, 5/8, and 7/8, so $n^{(1)}(3) = 4$; and in general,

$$n^{(1)}(k) = \begin{cases} 2 & k = 0 \\ 2^{k-1} & k > 0. \end{cases} \tag{4}$$

We now denote by $N^{(1)}(k)$ the number of distinct values in one dimension constructible with k or fewer folds. It is given by

$$N^{(1)}(k) = n^{(1)}(0) + n^{(1)}(1) + \cdots + n^{(1)}(k)$$
$$= 2 + 1 + 2 + 4 + \cdots + 2^{k-1}$$
$$= 1 + 2^k. \tag{5}$$

Now, to treat the two-dimensional case, we observe that we can obtain a unique point for every unique pair (x,y) where x and y are constructed by the binary method. With exactly k folds at our disposal, we can partition the k folds between x- and y-coordinate constructions; if we use i folds for x, then we have $k - i$ folds available for the y-coordinate. Thus, the number of points constructible with exactly k folds in two-dimensions, which we denote by $n^{(2)}(k)$, is given by $n^{(2)}(0) = 4$ and, for $k > 0$,

$$n^{(2)}(k) = n^{(1)}(0) \cdot n^{(1)}(k) + n^{(1)}(1) \cdot n^{(1)}(k-1) + \cdots + n^{(1)}(k) \cdot n^{(1)}(0)$$
$$= 2 \cdot 2^{k-1} + 1 \cdot 2^{k-2} + 2 \cdot 2^{k-3} + \cdots + 2^{k-2} \cdot 1 + 2^{k-1} \cdot 2$$
$$= 2^k + (k-1) \cdot 2^{k-2} + 2^k$$
$$= (k+7) \cdot 2^{k-2}. \tag{6}$$

As we did in one-dimension, we can compute $N^{(2)}(k)$, the number of distinct two-dimensional points constructible with k or fewer folds, by summing over $n^{(2)}(k)$:

$$N^{(2)}(k) = n^{(2)}(0) + n^{(2)}(1) + \cdots + n^{(2)}(k)$$
$$= 4 + 4 + \sum_{i=2}^{k} (i+7) \cdot 2^{i-2}$$
$$= 8 + \sum_{i=2}^{k} (i-2) \cdot 2^{i-2} + 9 \cdot \sum_{i=2}^{k} 2^{i-2} \tag{7}$$
$$= 8 + \sum_{i=0}^{k-2} i \cdot 2^i + 9 \cdot \sum_{i=0}^{k-2} 2^i$$
$$= 8 + \left[(k-1) \cdot 2^{k-1} - 2 \cdot 2^{k-1} + 2 \right] + \left[9 \cdot (2^{k-1} - 1) \right]$$
$$= 1 + (k-6) \cdot 2^{k-1}$$

The first seven values of $N^{(2)}(k)$ are given in Table 1.

k	$N^{(2)}(k)$
0	4
1	8
2	17
3	37
4	81
5	177
6	385

Table 1. The number of distinct points constructible with k or fewer creases by the binary method, for k up to 6.

Observe that the number of constructible points scales as $k \cdot 2^k$ for large k. The extra factor of k arises because the points thereby constructed are not uniformly distributed across the square. For example, for $k = 5$, the vertical line at $x = 1/2$ contains the 17 points constructible with four folds, while the vertical line at $x = 1/32$ only contains two points ($y = 0$ and $y = 1$).

If the points were uniformly distributed across the square, we would expect the average error to be something like $1/\left(2\sqrt{N^{(2)}(k)}\right)$, or about 0.026 for $k = 6$ (six folds). This is not terribly accurate, and the nonuniform distribution further increases the average error. However, there are more ways of creating points and lines by folding, and by considering all possible combinations, many more points—and thus, far more accurate approximations—are possible, as will be shown in the next section.

Axiomatic Origami

The binary approximation scheme discussed above created new lines by folding one line to another parallel line. However, there are several more ways that a fold line can be defined. For example, we can fold a point to another point, fold a line to another line (angle bisection), or put a crease through one or two points, to name a few. Starting in the 1970s, several folders began to systematically enumerate the possible combinations of folds and to study what types of distances were constructible by combining them in various ways. The first systematic study was carried out by Humiaki Huzita [3–5], who described a set of six basic ways of defining a single fold by aligning various combinations of existing points, lines, and

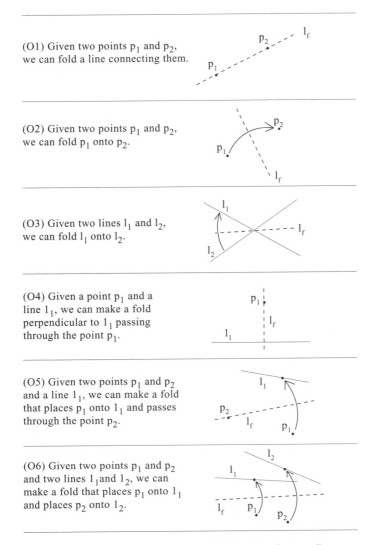

(O1) Given two points p_1 and p_2, we can fold a line connecting them.

(O2) Given two points p_1 and p_2, we can fold p_1 onto p_2.

(O3) Given two lines l_1 and l_2, we can fold l_1 onto l_2.

(O4) Given a point p_1 and a line l_1, we can make a fold perpendicular to l_1 passing through the point p_1.

(O5) Given two points p_1 and p_2 and a line l_1, we can make a fold that places p_1 onto l_1 and passes through the point p_2.

(O6) Given two points p_1 and p_2 and two lines l_1 and l_2, we can make a fold that places p_1 onto l_1 and places p_2 onto l_2.

Figure 6. The six operations of "Huzita's Axioms."

the fold line itself. These six operations have become known as "Huzita's Axioms" (HA). Given a set of points and lines on a sheet of paper, Huzita's operations allow one to create new lines; the intersections among old and new lines define additional points. The expanded set of points and lines may then be further expanded by repeated application of the operations to obtain further combinations of points and lines.

(O7) Given a points p_1 and two
lines l_1 and l_2, we can make a fold
perpendicular to l_2 that places p_1
onto line l_1.

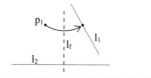

Figure 7. Hatori's seventh axiom.

An excellent introduction to "Huzita's Axioms" is given by Hull in [6], and I adopt his notation here. The six operations identified by Huzita are shown in Figure 6.

As with compass and straightedge, operations O1–O5 can be used to construct the solution of any quadratic equation with rational coefficients. Operation O6 is unique in that it allows the construction of solutions to the general cubic equation.

Recently, a seventh operation was proposed by Hatori [7], which I will denote by O7. It is shown in Figure 7.

Hatori noted that this operation was not equivalent to any of HA. Hatori's O7 allows the solution of certain quadratic equations (equivalently, it can be constructed by compass and straightedge). If we denote the expanded set as the "Huzita-Hatori operations" (HH operations), it can be shown that this set is complete, that is, these are all of the operations that define a single fold by alignment of combinations of points and finite line segments.

Approximation by Computer

As we have seen, it is possible to approximate any point within a square using successive bisection of the two sides, i.e., binary approximation. However, constructing the x- and y-coordinates independently is a rather inefficient method of locating a point. The binary algorithm for locating a point only makes use of one of the seven HH operations, specifically O2 ("given two points p_1 and p_2, we can fold p_1 onto p_2"), and then only considers points along a single edge of the square. By allowing the use of all seven HH operations and all previously-constructed points, it is possible to construct many more points of a given rank, permitting much higher levels of accuracy of approximation.

The question then becomes: what are the constructible points of a given rank? This question can be addressed by recursively constructing all possible points.

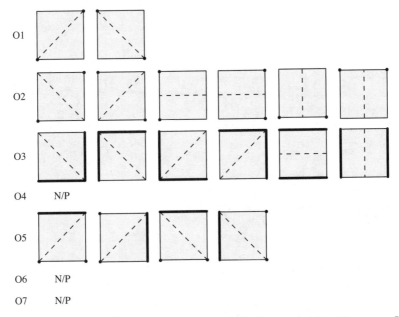

O1

O2

O3

O4   N/P

O5

O6   N/P

O7   N/P

Figure 8. The constructible lines on an unmarked square using the seven HH operations. The points and lines involved in the construction are highlighted.

Consider first the case $k = 0$, i.e., an unmarked square. In this case, there are four identifiable points, which are the four corners, and four lines, which are the four edges. It takes no folds to identify the corners of the square; we therefore assign the four corners a rank of zero.

We can also assign a rank to a fold line as well as to a point; the rank of a line is the number of folds it takes to create the line. In an unfolded square, there are four identifiable lines, which are the four edges of the square. Since they take no folds to construct, these four lines get a rank of zero as well. Thus, a square has four distinct points and four distinct lines of rank $k = 0$.

Now consider $k = 1$. The possible folds for each operation are illustrated in Figure 8 for O1, O2, O3, and O5. Operations O4, O6, and O7 do not (yet) permit the creation of any new lines.

Figure 8 shows that among all HH operations, there are four distinct new lines that can be created: the two diagonals and the two midlines of the square. Since each of these lines requires one fold to create, each has rank 1.

The intersections between the four new lines with each other and with the original edges of the square define five new points: the midpoints of

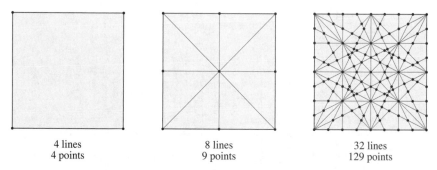

4 lines	8 lines	32 lines
4 points	9 points	129 points

Figure 9. Constructible points and lines using operations O1 and O2. Left: the initial square defines four lines and four points. Middle: all combinations of those define a total of eight lines and nine points. Right: all combinations of those define 32 lines and 129 points.

the sides and the very center of the square. Each new point along the edge of the square is defined by the intersection of a rank-0 edge and a rank-1 line; since each can be formed with a single fold, we therefore give them all a rank of 1. The center point may be defined as the intersection of several pairs of lines, but in all combinations, both lines are rank-1; therefore, the center point is rank-2. There are now a total of eight distinct lines and nine distinct points, eight of which have rank $k \leq 1$.

Now, let us consider making one more fold. With nine distinct points, there are

$$\binom{9}{2} = 36$$

possible pairs of points. Each of operations O1 and O2 acts on a pair of points and creates a new line, so there are 72 potential new lines. Some of these are duplicates of each other or of lines previously created, but 24 of them are distinct, bringing the number of lines to a total of 32 after two folds. Every intersection between two of the 32 lines potentially creates another point. After weeding out duplicates, in this second-stage of construction using just operations O1 and O2, we can construct a total of 32 distinct lines and 129 distinct points, as illustrated in Figure 9.

We still have five more operations to consider. Operation O3 acts on pairs of lines; O4 acts on point-line combinations; O5 acts on two points and a line; and so forth. Each operation creates geometrically more lines, pairs of whose intersections define geometrically more points. Even though duplications are inevitable, the number of distinct possible lines and points increases exponentially with the number of allowed folds.

As we make more folds, the rank of the newly created folds and points can be expressed in terms of the ranks of the points and lines that are brought into alignment to create them. A new point is always defined as the intersection of exactly two lines, and its rank is given by

$$k_p = k_{l_1} + k_{l_2}. \tag{8}$$

A new fold line l_f can be created in several ways by combining existing points $\{p_i\}$ and lines $\{l_j\}$, and the resulting rank is one more than the sum of the ranks of the points and lines used in its construction:

$$k_{l_f} = 1 + \sum_i k_{p_i} + \sum_j k_{l_j}. \tag{9}$$

If we open up the acceptable operations to include all seven HH operations, the combinatorics explode. Simply counting up the number of ways of combining points and lines among all possible operations gives the sequence $N = \{4, 258, 154{,}800, 132{,}826{,}269, \ldots\}$, which grows by about a factor of 1,000 with each iteration. However, only a portion of the possible combinations are physically realizable, and those include many duplicates— identical points that can be constructed with different folding sequences. The number of *distinct* constructible points is far smaller than the combinatorial limit.

A practical difficulty is that knowing that a simply-constructible point is somewhere near a target point is not the same as knowing what the nearest constructible point actually *is*. It would be nice if, given an arbitrary point (x, y), we could find a formula for the nearest constructible point of a given rank and the folding sequence for its construction. For the general case, we allow all of the HH operations, and allow any combination of lines and points that creates a line within the square. Unfortunately, for the general case, there is no known method for efficiently finding the nearest constructible point of a given rank, and I strongly suspect that no such method exists.

Fortunately, even inefficient methods can be suitable for practical application. Since 10^6 points should suffice to provide an accuracy of around 0.0005, it would suffice to construct the 10^6 or so lowest-rank points and lines; then given a desired target point, one simply searches through them all to find the closest point. Obviously, this is not something that one does by hand; but it is quite possible for a computer.

I wrote a C++ program called *ReferenceFinder* that does just this [8]. It takes as input the coordinates of a target reference point and prints out the best folding sequences for locating that point. In its initialization, *ReferenceFinder* constructs a database of about 300,000 distinct lines and

Percentile	Error
10th	0.0004
20th	0.0006
50th	0.0013
80th	0.0024
90th	0.0032
95th	0.0042
99th	0.0081

Table 2. Percentile and error for sequences taken from 277,546 6-fold constructions of distinct points.

points of rank 6 or below, by recursively building up higher-rank points from the lower ranks, weeding out duplicates, near-duplicates, and combinations that are not physically realizable along the way. This fairly restrictive filtering results in a much more modest, but still impressive rate of growth in numbers of constructible points, which runs $N = \{4, 8, 65, 1033, 7009, 32{,}469, 277{,}546\}$.

Using the 277,546 points with rank of 6 or less, I picked 1,000 random target points, found the closest constructible points, and computed statistics on the distribution of errors. The results are shown in Table 2.

In general, an error of 0.005—1.2 mm out of a 25 cm square—is barely noticeable. For 97% of the target points, there is a 6-fold sequence that achieves that level of error. Compare that with the binary method, which requires 18 folds to achieve the same accuracy.

The improved accuracy arises from the fact that at each stage of the construction, the number of possible distinct creases and points is based on several different possible combinations of lower-rank objects, which leads to exponential growth; the exponential scaling constant is roughly related to the number of different ways that points and lines can be combined to yield new ones.

Computer solutions for efficient folding sequences are of more than academic interest. As origami designers turn to mathematical methods of designing origami, it becomes necessary to develop efficient folding sequences for reference points that are defined solely as the solution of high-order algebraic equations. Programs like *ReferenceFinder* can construct those folding sequences, which can be surprising in their efficiency. Several recent origami books [9–11] have incorporated such computer-generated folding sequences as part of the instruction of individual figures, and I anticipate that such usage will become more common in the future.

References

[1] James Brunton, "Mathematical exercises in paper folding," *Mathematics in School*, Longmans for the Mathematical Association, vol. 2, no. 4, July 1973, p. 25.

[2] Robert J. Lang, "Four Problems III," *British Origami*, no. 132, October, 1988, pp. 7–11.

[3] Humiaki Huzita and Benedetto Scimemi, "The Algebra of Paper-Folding (Origami)," *Proceedings of the First International Meeting of Origami Science and Technology*, H. Huzita, ed., University of Padova, Padova, 1989, pp. 215–222.

[4] Humiaki Huzita, "Understanding Geometry through Origami Axioms," *Proceedings of the First International Conference on Origami in Education and Therapy (COET91)*, J. Smith ed., British Origami Society, London, 1992, pp. 37–70.

[5] Thomas Hull, "Geometric Constructions via Origami," *Proceedings of the Second International Conference on Origami in Education and Therapy (COET95)*, V'Ann Cornelius, ed., Origami USA, New York, 1995, pp. 31–38.

[6] Thomas Hull, http://web.merrimack.edu/hullt/geoconst.html, 2003.

[7] Koshiro Hatori, http://www.jade.dti.ne.jp/~hatori/library/conste.html, 2003.

[8] Robert J. Lang, http://origami.kvi.nl/programs/reffind/index.htm, 2003.

[9] Robert J. Lang, *Origami Insects II*, Gallery Origami House, Tokyo, 2003.

[10] John Montroll, *A Plethora of Polyhedra in Origami,* Dover Publications, New York, 2002.

[11] Robert J. Lang, *Origami Design Secrets*, A K Peters, Ltd., Natick MA, 2003.

Rolling Block Mazes

M. Oskar van Deventer

Puzzles involving rolling cubes have existed for quite some time. A classic puzzle asks to roll a six-sided die over all 64 squares of a checkerboard exactly once without the six ever appearing on top except at the starting position in the top-left and the ending position in the top-right [8]. In autumn 1998, while attending a conference in Belgium, I developed my

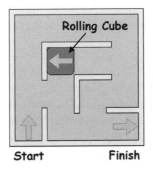

Start   **Finish**

Figure 1. Belgian Maze: Can you roll the cube from start to finish, at both ends having the arrow on the top face with the specified orientation?

M. Oskar van Deventer is the creator of hundreds of innovative mechanical puzzle designs, several of which are commercially available.

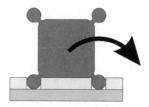

Figure 2. Rolling cube mechanism for the Belgian Maze.

first rolling block maze. Later, I learned that Richard Tucker invented the concept of a rolling block maze exactly at the same time; see [1].

My first rolling cube maze design was called Belgian Maze. The puzzle consists of a simple 4 × 4 maze, shown in Figure 1, and a cube having one face labeled with an arrow. The cube must start with the arrow on the top face with the orientation specified by the maze. The goal is to reach the finish with the arrow again on the top face and matching the orientation specified by the maze. While the 4 × 4 maze looks easy, the requirements on the arrow make it surprisingly challenging.

Figure 2 shows an effective physical mechanism I found to enforce the rolling movement of the cube, in which each corner of the cube has a ball-shape extrusion and each corner of a square in the maze has a matching indentation.

At the 1999 International Puzzle Party in London, I showed my design to Bill Ritchie, president of the Binary Arts corporation. Bill brought Adrian Fisher and me together, as Adrian was also working on rolling cube mazes. Adrian had developed a lot of rolling cube maze challenges on paper, based on the work of Robert Abbott [2]. Adrian had come up with the idea of the "rolling tea chest." The open side of the tea chest should never be down as then the precious china would fall out. Another idea of Adrian was the horse-jumping move. Two cubes are glued together and roll over a square grid. The piece could "jump" over barriers.

Combining my mechanism and Adrian Fisher's ideas, I developed the Rolling Block Maze shown in Figure 3. The barrier notches can be snapped easily onto the 5 × 5 grid to have a great variety of challenges. There are three pieces:

1. a simple rolling cube (shown in the lower-right),

2. a rolling cube with an antenna (shown in the middle-right), and

3. Adrian Fisher's jumping horse piece (shown in the upper-left).

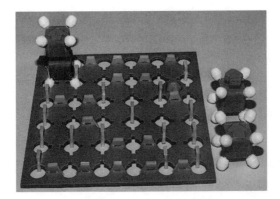

Figure 3. Rolling Block Maze.

The puzzle is much fun to play with, but there are some practical problems. The three rolling pieces have a rather complex shape that is hard to manufacture. There are too many small loose pieces. Moreover, it takes a lot of time to place the barriers and set up a challenge.

In early 2001, I succeeded in improving the mechanism by radically simplifying it. The result was the Pathfinder puzzle (named after the NASA Mars Pathfinder microrover, which it somewhat resembles). The rolling piece is made by gluing ten white beads into a $2 \times 3 \times 2$ "U"-shape; see Figure 4. Additional beads serve as obstacles. The shape is rolled over a 7×7 grid with indentations. Many different challenges are possible by placing the obstacle beads; see for example Figure 5.

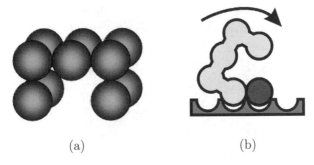

(a)   (b)

Figure 4. (a) Pathfinder shape; (b) rolling over an obstacle bead.

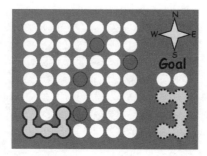

Figure 5. Pathfinder Challenge: Can you roll the Pathfinder from the configuration in the lower-left to the configuration in the lower-right, avoiding the shaded obstacles and missing indentations?

I made one hundred samples of Pathfinder as my exchange gift for the 2002 International Puzzle Party. It was quite a job to glue one thousand beads together. Sjaak Griffioen arranged the grid, made by vacuum forming; see Figure 6.

While working on Pathfinder, two people suggested alternative mechanisms. Wei-Hwa Huang proposed a cube frame (consisting of just the edges of a cube) that rolls over a grid of square-base pyramids. Adrian Fisher suggested removing four edges from the cube frame to make a three-dimensional "C"-shape, because then it can be made by one continuous path of bent wire. As shown in Figure 7, the frame can also be extended into a rolling "chair" shape or a rolling "S"-shape. Unfortunately, the piece does not roll very smoothly, the puzzle looks complex, and the resulting challenges are not too interesting.

Figure 6. Pathfinder as mechanical maze puzzle. Only half of the rolling piece (with five glued beads instead of ten) is in the lower-right.

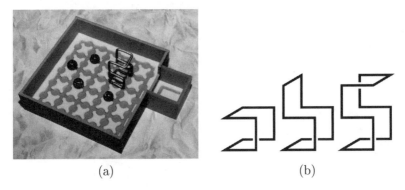

(a)   (b)

Figure 7. (a) Mechanical implementation of rolling "S"; (b) Rolling "C," rolling chair, and rolling "S."

Several people explored the whole pentomino alphabet to find alternative rolling shapes; see [1].

- Adrian Fisher: Rolling Piano, Rolling Slab, Chasm Maze ("jumping horse")

- Robert Abbott: Rolling-Block Maze, Other-End-Up Maze

- Erich Friedman: Rolling Slab, U Maze, Ramp Maze, Multi-Level Maze

- Richard Tucker: Hayling Island Maze, Minimal Maze

- Andrea Gilbert: Color Zone Maze

My exploration of rolling block mazes has resulted in several other puzzles. So far I have only shown you designs that involve cubic shapes rolling over a plane. But there are many alternatives.

Why should the grid be square? Wei-Hwa Huang presented Rolling Rhomboid [3], which uses a triangular grid. Another example with a triangular grid is my TetraRoll, shown in Figure 8, which is a variation of Pathfinder. The rolling piece is made by gluing seven beads in three vertical levels: two on the bottom, four in the middle, and one on top. Again the piece rolls on the board with spherical indentations.

Why should the grid be regular? My Rolling Cuboctahedron, shown in Figure 9, rolls over an irregular grid of triangles and squares. Here the rolling piece is a cuboctahedron, and the piece can be rolled around its edges in any way such that a cuboctahedron face always comes in full contact with a grid cell. Because neighboring faces on the octahedron are

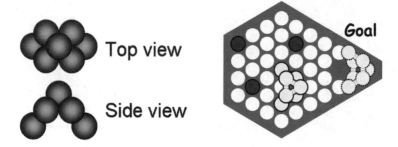

Figure 8. TetraRoll featuring a triangular grid. The drawing shows the starting position of a challenge. The goal is to roll the piece all the way to the right.

always one triangle and one square, this puzzle can also be presented as a maze of walking from the start cell to the finish cell with the restriction of passing alternately between squares and triangles.

Why should the rolling piece be regular? In my Rolling Turned Cuboctahedron, shown in Figure 10, the rolling piece is formed by cutting a cuboctahedron in half along a plane that passes through six edges, turning one half 60° about the centroid in that plane, and re-attaching the two halves. This "turned cuboctahedron" is the 27th Johnson solid, normally called the triangular orthobicupola. The maze on which the piece rolls is a regular grid of triangles and squares. There are several ways to roll from

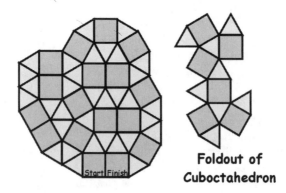

Foldout of Cuboctahedron

Figure 9. Rolling Cuboctahedron featuring an irregular grid of triangles and squares. Can you roll a cuboctahedron from start to finish, alternating between square and triangle?

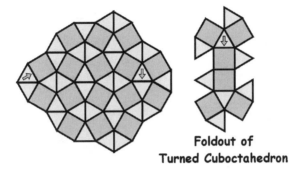

**Foldout of
Turned Cuboctahedron**

Figure 10. Rolling Turned Cuboctahedron featuring an irregular rolling shape. Can you roll the turned cuboctahedron from start (on the left) to finish (near the right), at both locations having the arrow on the top face with the specified orientation?

start to finish. However, if we label one triangular side of the rolling piece with an arrow, and start with that arrow on the top side in the specified orientation, there is only one way to roll to the finish with the arrow again showing on the top side in the specified orientation.

Why should the grid tile a region of the plane without holes? My Rolling Dodecahedron #I, shown in Figure 11, has a grid with pentagons

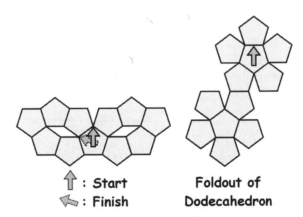

⬆ : Start
⬅ : Finish

**Foldout of
Dodecahedron**

Figure 11. Rolling Dodecahedron #I featuring a grid with holes. Can you roll the dodahedron from the central cell back, keeping the arrow on the top face but changing the orientation as specified?

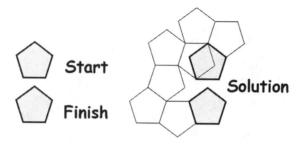

Figure 12. Rolling Dodecahedron #II has no grid at all.

and holes between them, The object is to roll a regular dodecahedron in such a way that the arrow gets turned by 72°.

Why should there be a grid anyway? In my Rolling Dodecahedron #II, shown in Figure 12, the start and finish positions for a dodecahedron seem to be vertically offset by an arbitrarily amount. In fact, the positions are constructed by considering the solution shown on the right of Figure 12. This puzzle is less practical to implement mechnically because it relies on exact rolling without guiding indentations.

Why should the grid be two-dimensional? In my DeltaRoll, shown in Figure 13, there are two pieces made of Velcro that roll around each other. The piece on the left is the 14-sided deltahedron (convex polyhedron whose faces are all equilateral triangles) resulting from augmenting a triangular prism by adding a square pyramid along each of the three square faces. This deltahedron is therefore called the triaugmented triangular prism, also known as the 51st Johnson solid. The piece on the right is the 12-sided deltahedron called the snub disphenoid or 84th Johnson solid. This deltahedron can be produced by cutting two triangular dipyramids each along a path of two edges connecting opposite apices of the dipyramid,

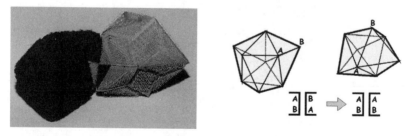

Figure 13. DeltaRoll, rolling in three dimensions.

Figure 14. Velcro Blocks have multiple challenges.

opening up the cuts, and joining the two open cuts with a half twist. Each piece is predominantly of one parity of Velcro, except near one edge where it is part of each parity. In the initial configuration (as shown on the left of Figure 13), the two pieces still meet with opposite-parity Velcro and therefore stick to each other. The goal of the puzzle is to reverse the orientation of one polyhedron so that the same parities of Velcro come together and the pieces fall apart.

Why should there be a single challenge? My Velcro Blocks, shown in Figure 14, consists of two pieces that roll around each other, a $1 \times 1 \times 2$ block and a $1 \times 2 \times 2$ block. In addition, each piece has triangular bumps and matching triangular indentations at certain locations to prevent certain moves. There are four start positions and four finish positions, specified using the colors of the bumps. (Velcro makes a ripping sound when you separate two parts. It is an ideal material for making things that roll.)

Finally, why should there be a single rolling block? Erich Friedman uses multiple rolling blocks in his Multi-Level Maze [7]. Dieter Gebhardt presented an overview of multiple-rolling-cube puzzles [4]. Serhiy Grabarchuk designed Two Real Dice, a combinination of rolling cube and sliding piece puzzles [5]. One could imagine rolling block puzzles with many pentomino-like blocks that operate in a similar way as Edward Hordern's sliding piece puzzles [6], but with rolling blocks instead of sliding pieces. Will this be the next type of Rolling Block Puzzle to be explored?

Solutions to Puzzles

Classic die rolling puzzle. E, E, E, S, W, S, W, N, W, S, S, S, S, S, S, E, N, E, S, E, N, N, W, W, N, N, E, S, E, N, N, E, S, S, E, N, E, S, S, W, W, S, S, E, N, E, S, E, N, N, N, N, N, N, W, S, W, N, W, N, E, E, E.

Figure 1. N, N, N, E, E, E, S, W, W, S, S, E, N, E, N, W, W, S, W, N, N, E, E, E, S, S, W, S, E.

Figure 5. N, N, N, N, E, E, N, W, W, S, E, S, E, E, S, W, S, E, N, N, N, W, W, W, S, E, S, W, S, E, E, E, E, N, E, E, S.

Figure 9. N, NW, NE, NW, NE, N, N, NW, NE, E, E, SE, SW, S, E, SE, SW, S, S, SE, SW, W.

Figure 10. NE, SE, E, NE, SE, NE, SE, E, SE, NE, SE, NE, NW, SW.

Figure 11. Roll clockwise around left circuit, then roll clockwise around right circuit

References

[1] Robert Abbott, "Logicmazes", internet website http://www.logicmazes.com/.

[2] Robert Abbott, "A Maze with Rules". In *The Mathemagician and Pied Puzzler*, edited by Elwyn Berlekamp and Tom Rodgers. Natick, MA: A K Peters, 1999, pp.27-28.

[3] Ed Pegg, "Mathpuzzle", internet website http://www.mathpuzzle.com/.

[4] Dieter Gebhardt, "Rolling Cube Puzzles". *Cubism For Fun*, no. 56, October 2001, pp.6-10.

[5] Serhiy Grabarchuk, "Two Real Dice", IPP20 exchange gift, Los Angeles, 2000.

[6] L.E, Hordern, *Sliding Piece Puzzles*. Recreations in Mathematics, No. 4. Oxford: Oxford University Press, 1986.

[7] Erich Friedman, "Rolling Block Mazes", internet website http://www.stetson.edu/~efriedma/rolling/.

[8] Miodrag Petković, *Mathematics and Chess*. New York: Dover, 1997, pages 95 and 102.

Constructing Domino Portraits

Robert Bosch

Introduction

In 1993 the artist Ken Knowlton created a portrait of Martin Gardner out of six complete sets of double nine dominoes; he presented the portrait to Gardner at G4G1, the first "Gathering for Gardner" conference [1]. Figure 1 displays domino portraits of Marilyn Monroe and John Lennon that were made by this author out of nine complete sets (per portrait) of double nine dominoes. In this brief note, we describe a new, integer-programming-based method for constructing approximations of target images using complete sets of double nine dominoes.

Dominoes

Figure 2 displays a complete set of double nine dominoes. There are 55 dominoes in a complete set—ten doubles and $\binom{10}{2} = 45$ non-doubles. Each

Robert Bosch teaches mathematics at Oberlin College. He specializes in nontraditional application of optimization.

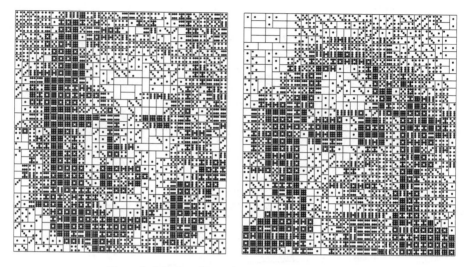

Figure 1. Marilyn (9 sets) and John (9 sets).

double has two orientations: v (vertical) and h (horizontal). Each non-double has four orientations: v_1 (vertical with the lower-numbered square on top), v_2 (vertical with the lower-numbered square on the bottom), h_1 (horizontal with the lower-numbered square on the left), and h_2 (horizontal with the lower-numbered square on the right).

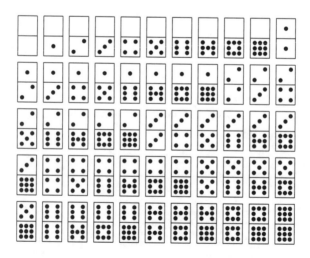

Figure 2. A complete set of double nine dominoes.

Because each domino is made of two squares, s^2 complete sets can be used to create pictures on an $11s \times 10s$ canvas of squares.

The Target Image

We first convert the target image into PGM (portable graymap) format. Each pixel of the image is given a grayscale value between 0 (completely black) and 255 (completely white). We then divide the image into $11s$ rows and $10s$ columns of $k \times k$ clusters of pixels, one cluster for each square of a domino in our canvas. Finally, for each cluster, we compute the mean grayscale value and then convert the mean to an integer between 0 (completely white) and 9 (completely black). We then let $g_{i,j}$ denote the resulting grayscale value of the image's square at row i and column j.

An Integer Programming Formulation

We formulate the problem of constructing an "ideal" domino portrait of the desired image as an integer program. (See [3] for a standard reference on integer programming.) Basically, an integer program is an optimization problem involving several integer variables, several linear equality or inequality constraints over the variables, and a linear objective function over the variables. The goal is to choose values for the variables in order to minimize or maximize the objective function subject to satisfying the constraints.

Decision Variables

We let $x_{m,n,o,i,j}$ equal 1 if domino (m, n) is placed in orientation o with its top left square in the row-i-column-j square of the canvas and 0 if not. (We keep $m \leq n$ throughout.) It is easy to show that if the canvas has $11s$ rows and $10s$ columns, then the number of decision variables is $22{,}000s^2 - 2{,}100s$.

Objective Function

We let $c_{m,n,o,i,j}$ equal the "cost" of placing domino (m, n) in orientation o with its top left square in the row-i-column-j square of the canvas. We measure costs using a 2-norm. For example, if $o = v_1$, then $c_{m,n,o,i,j} = (m - g_{i,j})^2 + (n - g_{i+1,j})^2$, the sum of penalties we charge ourselves for placing the "m" in square (i, j) and the "n" in square $(i + 1, j)$. Clearly, our goal is to minimize the total penalty

$$\sum_{m,n,o,i,j} c_{m,n,o,i,j} x_{m,n,o,i,j}.$$

Constraints

We need two types of constraints. The "type-one" constraint

$$\sum_{o,i,j} x_{m,n,o,i,j} = s^2$$

stipulates that domino (m, n) is to be used s^2 times. (Recall that we are using s^2 sets of dominoes.) We need 55 type-one constraints—one for each distinguishable domino. The "type-two" constraint

$$\sum_m x_{m,m,v,i,j} + \sum_m x_{m,m,v,i-1,j}$$
$$+ \sum_m x_{m,m,h,i,j} + \sum_m x_{m,m,h,i,j-1}$$
$$+ \sum_{m<n} x_{m,n,v_1,i,j} + \sum_{m<n} x_{m,n,v_2,i,j}$$
$$+ \sum_{m<n} x_{m,n,v_1,i-1,j} + \sum_{m<n} x_{m,n,v_2,i-1,j}$$
$$+ \sum_{m<n} x_{m,n,h_1,i,j} + \sum_{m<n} x_{m,n,h_2,i,j}$$
$$+ \sum_{m<n} x_{m,n,h_1,i,j-1} + \sum_{m<n} x_{m,n,h_2,i,j-1} = 1$$

states that the row-i-column-j square of the canvas must be covered by exactly one domino. Literally, it states that the row-i-column-j square is covered by a double domino placed vertically, or a double domino placed horizontally, or a non-double domino placed vertically, or a non-double domino placed horizontally. We need $110s^2$ type-two constraints—one for each square of the canvas.

Results

The resulting integer programs are quite large. Each "$s = 3$" (9-set) integer program has 191,700 decision variables and 1,045 constraints. Each

"$s = 7$" (49-set) integer program has 1,063,300 decision variables and 5,445 constraints.

Large integer programs are often intractable. Fortunately, these integer programs are usually very easy to solve. The integer programs used to create the 9-set portraits of Marilyn Monroe and John Lennon (see Figure 1) required 763 seconds and 567 seconds, respectively, and the integer program used to create the 25-set approximation of Vermeer's "Girl with a Pearl Earring" (see Figure 3) required 3,916 seconds. (All computations were performed with CPLEX 6.6 on an 800 Mz Pentium III PC.) The integer program used to create a 49-set approximation (not shown) of the same Vermeer painting required 15,816 seconds. See [2] for more pictures of domino portraits that were constructed with integer programming, some built out of real dominoes.

Figure 3. Vermeer's Girl with a Pearl Earring (25 sets).

References

[1] E. Berlekamp and T. Rodgers. *The Mathemagician and Pied Puzzler: A Collection in Tribute to Martin Gardner.* Natick, MA: A. K. Peters, 1999.

[2] R. Bosch. *Domino Artwork.* Available at http://www.dominoartwork.com, 2002.

[3] A. Schrijver. *Theory of Linear and Integer Programming.* New York: John Wiley & Sons, 1998.

Index

by Author

So far three books—*The Mathemagician and Pied Puzzler* (edited by Elwyn Berlekamp and Tom Rodgers), *Puzzlers' Tribute: A Feast for the Mind* (edited by David Wolfe and Tom Rodgers), and *Tribute to a Mathemagician*—have been published to honor Martin Gardner and to compile contributions made at the G4G gatherings. *The Mathemagician and Pied Puzzler* includes contributions from G4G1. *Puzzlers' Tribute: A Feast for the Mind* is a compilation from G4G2 to G4G4. Contributions from G4G5 appear in *Tribute to a Mathemagician*.

Included here is an index that references the authors of all three books. Entries are read:

Author Name. *Article Title.* (Book Number) Page Number

where the Book Number (1, 2, or 3) refers to

(1) *The Mathemagician and Pied Puzzler*

(2) *Puzzlers' Tribute: A Feast for the Mind*

(3) *Tribute to a Mathemagician*